Applied and Numerical Harmonic Analysis

Series Editor
John J. Benedetto
University of Maryland
College Park, MD, USA

Editorial Advisory Board

Akram Aldroubi
Vanderbilt University
Nashville, TN, USA

Douglas Cochran
Arizona State University
Phoenix, AZ, USA

Hans G. Feichtinger
University of Vienna
Vienna, Austria

Christopher Heil
Georgia Institute of Technology
Atlanta, GA, USA

Stéphane Jaffard
University of Paris XII
Paris, France

Jelena Kovačević
Carnegie Mellon University
Pittsburgh, PA, USA

Gitta Kutyniok
Technische Universität Berlin
Berlin, Germany

Mauro Maggioni
Johns Hopkins University
Baltimore, MD, USA

Zuowei Shen
National University of Singapore
Singapore, Singapore

Thomas Strohmer
University of California
Davis, CA, USA

Yang Wang
Hong Kong University of Science & Technology
Kowloon, Hong Kong

For further volumes:
http://www.springer.com/series/4968

Elijah Liflyand

Functions of Bounded Variation and Their Fourier Transforms

Elijah Liflyand
Department of Mathematics
Bar-Ilan University
Ramat-Gan, Israel

ISSN 2296-5009   ISSN 2296-5017 (electronic)
Applied and Numerical Harmonic Analysis
ISBN 978-3-030-04428-2   ISBN 978-3-030-04429-9 (eBook)
https://doi.org/10.1007/978-3-030-04429-9

Library of Congress Control Number: 2018968411

Mathematics Subject Classification (2010): 42A38, 42A32, 42A50, 42B10, 42B05, 42B30, 42B35, 26A45, 26A46, 26B30

© Springer Nature Switzerland AG 2019
This work is subject to copyright. All rights are reserved by the Publisher, whether the whole or part of the material is concerned, specifically the rights of translation, reprinting, reuse of illustrations, recitation, broadcasting, reproduction on microfilms or in any other physical way, and transmission or information storage and retrieval, electronic adaptation, computer software, or by similar or dissimilar methodology now known or hereafter developed.
The use of general descriptive names, registered names, trademarks, service marks, etc. in this publication does not imply, even in the absence of a specific statement, that such names are exempt from the relevant protective laws and regulations and therefore free for general use.
The publisher, the authors and the editors are safe to assume that the advice and information in this book are believed to be true and accurate at the date of publication. Neither the publisher nor the authors or the editors give a warranty, express or implied, with respect to the material contained herein or for any errors or omissions that may have been made. The publisher remains neutral with regard to jurisdictional claims in published maps and institutional affiliations.

This book is published under the imprint Birkhäuser, www.birkhauser-science.com by the registered company Springer Nature Switzerland AG.
The registered company address is: Gewerbestrasse 11, 6330 Cham, Switzerland

Those who we learn from are called teachers...

Johann Wolfgang von Goethe

TO MY TEACHERS

ANHA Series Preface

The *Applied and Numerical Harmonic Analysis (ANHA)* book series aims to provide the engineering, mathematical, and scientific communities with significant developments in harmonic analysis, ranging from abstract harmonic analysis to basic applications. The title of the series reflects the importance of applications and numerical implementation, but richness and relevance of applications and implementation depend fundamentally on the structure and depth of theoretical underpinnings. Thus, from our point of view, the interleaving of theory and applications and their creative symbiotic evolution is axiomatic.

Harmonic analysis is a wellspring of ideas and applicability that has flourished, developed, and deepened over time within many disciplines and by means of creative cross-fertilization with diverse areas. The intricate and fundamental relationship between harmonic analysis and fields such as signal processing, partial differential equations (PDEs), and image processing is reflected in our state-of-the-art *ANHA* series.

Our vision of modern harmonic analysis includes mathematical areas such as wavelet theory, Banach algebras, classical Fourier analysis, time-frequency analysis, and fractal geometry, as well as the diverse topics that impinge on them.

For example, wavelet theory can be considered an appropriate tool to deal with some basic problems in digital signal processing, speech and image processing, geophysics, pattern recognition, biomedical engineering, and turbulence. These areas implement the latest technology from sampling methods on surfaces to fast algorithms and computer vision methods. The underlying mathematics of wavelet theory depends not only on classical Fourier analysis, but also on ideas from abstract harmonic analysis, including von Neumann algebras and the affine group. This leads to a study of the Heisenberg group and its relationship to Gabor systems, and of the metaplectic group for a meaningful interaction of signal decomposition methods. The unifying influence of wavelet theory in the aforementioned topics illustrates the justification for providing a means for centralizing and disseminating information from the broader, but still focused, area of harmonic analysis. This will be a key role of *ANHA*. We intend to publish with the scope and interaction that such a host of issues demands.

Along with our commitment to publish mathematically significant works at the frontiers of harmonic analysis, we have a comparably strong commitment to publish major advances in the following applicable topics in which harmonic analysis plays a substantial role:

Antenna theory
Biomedical signal processing
Digital signal processing
Fast algorithms
Gabor theory and applications
Image processing
Numerical partial differential equations
Prediction theory
Radar applications
Sampling theory
Spectral estimation
Speech processing
Time-frequency and time-scale analysis
Wavelet theory

The above point of view for the *ANHA* book series is inspired by the history of Fourier analysis itself, whose tentacles reach into so many fields.

In the last two centuries Fourier analysis has had a major impact on the development of mathematics, on the understanding of many engineering and scientific phenomena, and on the solution of some of the most important problems in mathematics and the sciences. Historically, Fourier series were developed in the analysis of some of the classical PDEs of mathematical physics; these series were used to solve such equations. In order to understand Fourier series and the kinds of solutions they could represent, some of the most basic notions of analysis were defined, e.g., the concept of "function." Since the coefficients of Fourier series are integrals, it is no surprise that Riemann integrals were conceived to deal with uniqueness properties of trigonometric series. Cantor's set theory was also developed because of such uniqueness questions.

A basic problem in Fourier analysis is to show how complicated phenomena, such as sound waves, can be described in terms of elementary harmonics. There are two aspects of this problem: first, to find, or even define properly, the harmonics or spectrum of a given phenomenon, e.g., the spectroscopy problem in optics; second, to determine which phenomena can be constructed from given classes of harmonics, as done, for example, by the mechanical synthesizers in tidal analysis.

Fourier analysis is also the natural setting for many other problems in engineering, mathematics, and the sciences. For example, Wiener's Tauberian theorem in Fourier analysis not only characterizes the behavior of the prime numbers, but also provides the proper notion of spectrum for phenomena such as white light; this latter process leads to the Fourier analysis associated with correlation functions in filtering and prediction problems, and these problems, in turn, deal naturally with Hardy spaces in the theory of complex variables.

Nowadays, some of the theory of PDEs has given way to the study of Fourier integral operators. Problems in antenna theory are studied in terms of unimodular trigonometric polynomials. Applications of Fourier analysis abound in signal processing, whether with the fast Fourier transform (FFT), or filter design, or the adaptive modeling inherent in time-frequency-scale methods such as wavelet theory. The coherent states of mathematical physics are translated and modulated Fourier transforms, and these are used, in conjunction with the uncertainty principle, for dealing with signal reconstruction in communications theory. We are back to the raison d'être of the *ANHA* series!

University of Maryland
College Park

John J. Benedetto
Series Editor

Contents

ANHA Series Preface   vii

Foreword   xv

Introduction   1

Part I: One-dimensional Case   9

1   A toolkit   11
- 1.1   Functions of bounded variation 11
- 1.2   Fourier transform 16
- 1.3   Hilbert transform 19
  - 1.3.1   Fourier transform weakly generates Hilbert transform . . . 21
  - 1.3.2   Existence almost everywhere 23
  - 1.3.3   Integrability of the Hilbert transform 28
  - 1.3.4   Special cases of the Hilbert transform 30
  - 1.3.5   Conditions for the integrability of the Hilbert transform . . 33
- 1.4   Hardy spaces and subspaces 35
  - 1.4.1   Atomic characterization 36
  - 1.4.2   Molecular characterization 39
  - 1.4.3   Integrability spaces 42
  - 1.4.4   A Paley–Wiener theorem 50
- 1.5   Balance integral operator 52

2   Functions with derivative in a Hardy space   57
- 2.1   First steps 58
- 2.2   Derivative in $H_o^1(\mathbb{R}_+)$ 61
- 2.3   Derivative in $H_e^1(\mathbb{R}_+)$ 65
- 2.4   Derivative in a subspace 70
- 2.5   Functions on the whole axis 77
- 2.6   Hardy–Littlewood theorem 78

3 Integrability spaces: wide, wider and widest — 85
- 3.1 Widest integrability spaces 85
- 3.2 The sine Fourier transform 87
- 3.3 Intermediate spaces 91
  - 3.3.1 Embeddings 91
  - 3.3.2 A counterexample 93
  - 3.3.3 Intermediate spaces between H_0^1 and H_Q^1 95
- 3.4 Fourier–Hardy type inequalities 96

4 Sharper results — 99
- 4.1 The Fourier transform of a convex function 100
  - 4.1.1 General representation of the Fourier transform 101
  - 4.1.2 Convex functions 104
- 4.2 Generalizations of Theorems 2.8 and 2.20 106
- 4.3 The sine Fourier transform revisited 108
- 4.4 A Szökefalvi-Nagy type theorem 111

Part II: Multi-dimensional Case — 115

5 A toolkit for several dimensions — 117
- 5.1 Indicator notation 117
- 5.2 Multidimensional variations 119
  - 5.2.1 Vitali's and Hardy's variations 119
  - 5.2.2 Tonelli's variation 121
- 5.3 Fourier transform 121
  - 5.3.1 L^1-theory 122
  - 5.3.2 L^2- and L^p-theory 123
  - 5.3.3 Poisson summation formula 124
- 5.4 Multidimensional spaces 125
- 5.5 Absolute continuity 128
- 5.6 Integration by parts 130

6 Integrability of the Fourier transforms — 133
- 6.1 Functions with derivatives in the Hardy type spaces 133
- 6.2 Hardy–Littlewood theorem 138
  - 6.2.1 Commutativity 138
  - 6.2.2 Conditions for absolute continuity 139
  - 6.2.3 Hardy–Littlewood type theorems 140

7 Sharp results — 143
- 7.1 Convexity type results 144
  - 7.1.1 Functions of convex type 146
- 7.2 Equalities 148
  - 7.2.1 (Even) more general cases 152

		7.2.2	The most general situation	154
	7.3	Szökefalvi-Nagy type theorem		156
		7.3.1	Auxiliary results	157
		7.3.2	Proof of Theorem 7.17	159

8 Bounded variation and sampling — 161
  8.1 Bridge — 161
   8.1.1 One-dimensional bridge — 163
   8.1.2 Temporary bridge — 166
   8.1.3 Stable bridge — 168
  8.2 On the Poisson summation formula — 172
   8.2.1 Background — 172
   8.2.2 A version of the Poisson summation formula — 173
   8.2.3 Concluding remarks and an example — 176

9 Multidimensional case: radial functions — 179
  9.1 Fractional derivative and MV classes — 179
  9.2 Necessary conditions — 180
  9.3 Radial extensions — 184

Afterword — 189

Bibliography — 193

Basic notations — 207

Index — 211

Foreword

It is rather predictable that many, just reading the title, will ask at least two questions: why, in fact, the functions of bounded variation and if so, why their Fourier transforms? The first question is redoubled with the fact that a book by J. Appell, J. Banas, and N. J. Merentes Diáz, *Bounded Variation and Around*, [9], has recently appeared. However, this book answers, in a sense, both questions. The point is that the present work can be considered as a supplement, a kind of the second volume of that book. The last chapter of that book, Chapter 7 entitled "Some applications" is extremely brief. This is by no means a criticism of the above three authors. Just the opposite, I am grateful to them for the room that they left for further work. The present book continues, so to say, from the point they ended on. More precisely, just the study of the Fourier transform of a function of bounded variation reveals numerous properties of such functions not enlightened or considered casually in [9]. These are the definition of the Fourier transform for functions of bounded variation, their various relations with the Hilbert transform and with the Hardy space, interesting spaces that appear during this study and their interrelations. Another objection might be that much of these is given in Chapter 3 of an even more recent book [92]. However, this objection can be withdrawn due to the surprising fact that during the last couple of years that chapter has needed to be rewritten almost completely. Not only new results have appeared but also new approaches to the old ones, including new proofs. And all this concerns the one-dimensional case. All the more so, new opportunities in the estimates of the remainder terms in one-dimensional asymptotic formulas allow one to obtain much more precise multidimensional generalizations.

Hoping that the above convinces the reader that this book ought to be written, let us discuss what it is about. More precisely, what point of view it gives on the outlined objects. A real target was not to present a collection of old and recent results, though this is partially done, especially concerning recent results, but to evolve a general picture of the topic, first of all in dimension one, but also in several dimensions. Of course, the latter is one of a variety of possible approaches, since one type of variation among many others, the so-called Hardy variation, is mainly touched in our study. There is a strong hope that this goal is achieved, at least, what is presented in this book could even not have been outlined a couple of years ago. However, there still are blind spots and open problems. Mathematically speaking, this affects simple connectivity of the picture but not connectivity. Moreover, it might be a merit of the topic and, correspondingly, of the book rather than a demerit. It shows that the subject is rich enough and leaves room for interesting problems and further study.

Since my student years I have had the privilege to learn immediately about new results of my teacher Professor Roald M. Trigub. My first result and its consequent publication came out as a multivariate extension of such a result unpublished at that time. The springboard for my interest in the problems of this book was Trigub's suggestion to generalize to the Fourier transform setting certain advanced results from the theory of integrability of trigonometric series. Later on, the first and the following results had taken a shape of the broader topic of which this book is about, perfectly fitting the saying "The best teachers are those who show you where to look, but don't tell you what to see" by that mysterious Alexandra K. Trenfor. Undoubtedly, certain progress and better understanding of these problems, if they exist, could not appear without both that first step and further continuous communications with Roald M. Trigub.

It is a real pleasure to express my sincere gratitude to other people that had an impact on my understanding of the problems considered within or related to the subject matter of this book: Rauan Akylzhanov, Luigi Ambrosio, Laura Angeloni, Eduard Belinsky Z"L, Roman Bessonov, Alexander Borichev, William Bray, Yuri Brudnyi, Paul L. Butzer, Laura De Carli, Sun-Yung Alice Chang, Michael Cwikel, Galia Dafni, Volodymyr Derkach, Seva Dyn'kin Z"L, Hans Feichtinger, Sandor Fridli, Michael Ganzburg, Anatoly Golberg, Dmitry Gorbachev, Victor Havin, Alex Iosevich, Raymond Johnson, Alexey Karapetyants, Dmitry Khavinson, Sergey Kislyakov, Paul Koosis, Natan Kruglyak, Olga Kuznetsova, Michael Lacey, Massimo Lanza de Cristoforis, Andrei Lerner, Guo Zhen Lu, Erwin Lutwak, Vladimir Matsaev Z"L, Ferenc Móricz, Fedor Nazarov, Rolf J. Nessel, Yuri Nosenko Z"L, Alexander Olevskii, Joan Orobitg, Jill Pipher, Isaac Pesenson, Anatoly Podkorytov, Simeon Reich, Fulvio Ricci, Mark Rudelson, Michael Ruzhansky, Stefan Samko, Eugene Shargorodsky, Igor Shevchuk, Maria Skopina, Mikhail Sodin, Boris Solomyak, Ulrich Stadtmüller, Alexander Stokolos, Sergey Tikhonov, Xavier Tolsa, Walter Trebels, Gianluca Vinti, Nicholas Young, Yuriy Zhaurov, and Georg Zimmermann. And last but not least, the author is also in debt to the anonymous referees for thorough reading and numerous suggestions.

This list apparently should be longer, since there is a high probability that I forgot some names. I apologize for this and definitely acknowledge the contribution of all my colleagues.

I would also like to thank Miriam Beller and Iryna Bykova for their assistance in the preparation of this manuscript

A Russian (living in Ukraine) writer of ironic genre, Vyacheslav Verkhovsky, mentioned once that "this book owes me the best in it". It was about his own book, of course, but I indulge in (proudly?) repeating this about the present text. However, there is a suspicion that the book also owes me the worst in it. It is the reader who weighs both aspects and either compliments or criticizes my work.

Elijah Liflyand

Introduction

Before starting this book, I indulge myself in citing a passage from our previous book [92]; moreover, the credit is due to my friend and coauthor of [92] Alex Iosevich:

The Fourier transform is a ubiquitous object in modern analysis, physics and engineering. Far from being a "trick" or a "tool", it is rather a fundamental operation which relates the spacial properties of a function with its frequency behavior.

In other words, the Fourier transform is a constitutive operation in harmonic analysis. Another fundamental object is the class of functions with bounded variation. In this book, we study these two objects in their interaction. Such a study sheds light on both of them. It turns out that frequently they are related via the other fundamental operation the Hilbert transform. Of course, this is not a novelty, one of the classical ways to see this is the multiplier-type formula

$$\widehat{\mathcal{H}g}(-x) = i\,\text{sign}x\,\widehat{g}(x).$$

However, there are much more points of contact of the afore-mentioned basic operations and objects. In a sense, all three of them control the oscillations of a function, each of them in its own way. Combinations of these actions lead not only to new results but also to new points of view, and all these is what this book is about.

Transforms

What about spacial properties in the present book? The main and most general one is that the function be of bounded variation. Instead of the general Fourier transform, we concentrate on the study of the cosine Fourier transform

$$\widehat{f_c}(x) = \int_0^\infty f(t)\cos xt\,dt$$

and the sine Fourier transform

$$\widehat{f_s}(x) = \int_0^\infty f(t)\sin xt\,dt,$$

and their integrability properties. Just so, separately. The difference between the two transforms is a known phenomenon. For instance, if a function is monotone on $\mathbb{R}_+ = [0,\infty)$, then its sine Fourier transform preserves the sign and furnishes more information (see Chapter 2). On the other hand the Fourier transform of such a function cannot be Lebesgue integrable on $\mathbb{R}$. For a convex function, the classical Pólya theorem brings the cosine transform to the fore. More precisely, the cosine Fourier transform of a bounded convex function, vanishing at infinity, preserves the sign and is Lebesgue integrable no matter how slow the function decays near infinity.

An interesting discussion on the interrelations between the sine and cosine transforms can be found in [202]:

There are multiple ways in which sine and cosine are related: as derivatives of each other, as distributional Hilbert transforms of each other and as linearly independent eigenfunctions of the Laplacian in one variable. The derivative operator does not play well with unitarity so this is not feasible; the distributional Hilbert transform, while rich with theory, is difficult to treat in practice. For these reasons, and more, we choose to use view sine and cosine as linearly independent solutions to the same differential equation.

We prefer to rely on the Hilbert transform, even not distributional but in the usual principal value sense. That the sine and cosine functions form a Hilbert transform pair plays a considerable role in this treatment. In fact, our functions are not less Hilbert-transformed than Fourier-transformed.

Of course, there will be digressions from the mentioned general line, for example, sometimes the Fourier transform will be considered for functions defined on the whole real axis and the whole second part will be devoted to multidimensional estimates, but all of them will be in one sense or another related to the outlined setting.

Functions

A variety of questions can be asked about the Fourier transform. One of them, crucial to our study, is whether the Fourier transform is integrable or not. The role of the integrable Fourier transform can hardly be underestimated. That the inverse Fourier formula is valid in this case is more than enough. Of course, this question is not less important for functions of bounded variation but it turns out that for such functions there are specific ways to clarify how far away from the integrability, if not integrable, the Fourier transform is. This knowledge is of importance for many applications, such as approximation, summability (Lebesgue constants), etc.

Sometimes all such functions are considered but more often the functions are from a subclass of the space of functions of bounded variation. What about their Fourier transforms, when they started to be considered, what is the store of knowledge about them on which the present book builds?

It can be said that only episodic attempts had been taken till recently that may be considered in the context of our study. Probably, the paper by Hille and Tamarkin [88] is the first, or first notable, one. Of course, Bochner's consideration of the Fourier transform of a monotone function in his celebrated monograph can be classed among those. The work of Sz.-Nagy in the 40-s is also of importance. Many known approaches and results for these functions were summarized in the well-known book by Butzer and Nessel.

It is safe to say that "modern history" of this area begins with Trigub's results on the Fourier transform of a convex function in the mid 70-s, which, in turn, generalize and refine earlier results by Shilov on the Fourier coefficients of a convex function. Later progress also was made by Trigub. It was the author then who, by a suggestion of Trigub, tried and succeeded to generalize the Boas–Telyakovskii-type theorem and certain related theorems. The work of Giang and Móricz should be mentioned as well.

Basic theorem

This is Theorem 2.8 in the book, a sort of point of departure. It is worth giving it in its explicit form immediately:

Let $f : \mathbb{R}_+ \to \mathbb{C}$ be locally absolutely continuous on $(0, \infty)$, of bounded variation and $\lim_{t\to\infty} f(t) = 0$. Assume also $f' \in H^1_o(\mathbb{R}_+)$. Then the cosine Fourier transform of f is integrable, with

$$\|\widehat{f}_c\|_{L^1(\mathbb{R}_+)} \lesssim \|f'\|_{H^1_o(\mathbb{R}_+)};$$

while for the sine Fourier transform an asymptotic formula holds: for $x > 0$,

$$\widehat{f}_s(x) = \frac{1}{x} f\left(\frac{\pi}{2x}\right) + F(x),$$

where

$$\|F\|_{L^1(\mathbb{R}_+)} \lesssim \|f'\|_{H^1_o(\mathbb{R}_+)}.$$

Here $H^1_o(\mathbb{R}_+)$ means the real Hardy space for odd functions, details are not important at this stage. What is more important already now is the class of functions to be considered in most situations, locally absolutely continuous on $(0, \infty)$, of bounded variation and vanishing at infinity, written $f \in BV_0[0, \infty) \cap LAC(0, \infty)$. For this class, the above theorem with its main essential assumption $f' \in H^1_o(\mathbb{R}_+)$ remained the main achievement till recently.

Why has it taken so long to advance from the basic Theorem 2.8 towards the recent refinements that will be outlined soon? Well, we know how fruitless work may be for ... ages until a sudden solution is discovered. That ideal "nulla dies sine linea" (Latin for "not a day without a line") is very often a dream or, even worse, leads to meaningless lines. Another Russian (and Israeli) ironic genre writer Igor Guberman wrote about this pretty vividly (translated jointly with Michael Cwikel):

> At my work desk from dawn till dark,
>
> eyes staring hard at paper; fine,
>
> what's left of the day? Not a spark,
>
> nor even a line.

If one wishes a more mathematical explanation, David Mumford's "The world is continuous, but the mind is discrete" copes with the task perfectly.

Advancement

However, then something happened, and new approaches started to appear along with consequent new results. The widest space for the integrability of the Fourier transform of a function of bounded variation came first, with a hint from a paper by Johnson and Warner, then there was an extension to amalgam-type spaces, then an asymptotic formula for the sine Fourier transform of **all**

$$f \in BV_0[0,\infty) \cap LAC(0,\infty),$$

then a counterpart of Theorem 2.8 with the real Hardy space for even functions, then more precise asymptotic formulas obtained with the help of a balance operator and new multidimensional extensions on the basis of the refined one-dimensional prototypes.

Among all the one-dimensional advancements given in the first part, let us give prominence to the new proof of Theorem 2.8 and Theorem 2.20, the proof of which uses the same idea. Their proofs as well as the proofs of certain of their corollaries proceed ultimately within the scope of the theory of Hardy spaces. One of the integrated parts of these proofs is Hardy's inequality

$$\int_{\mathbb{R}} \frac{|\widehat{g}(x)|}{|x|}\, dx \le \|g\|_{H^1(\mathbb{R})}.$$

In one form or another it worked in the previous approaches, though somewhat in the shadows, but now it is used directly and more actively. There is a linguistic problem with these tools: for many people Hardy's inequality has completely different meanings. Calling it Hardy–Littlewood's inequality, as some authors do, does not solve the problem, too many objects bear this double name. Our suggestion is to call it the *Fourier–Hardy inequality*, pointing out that the Fourier transform is essentially involved in the business. This terminology seems to have some basis and will be used throughout the book. Note that very recently, in one of the indicated works, an example has been given to confirm the non-triviality of the Fourier–Hardy inequality, that is, an example of a function not from the Hardy space but with finite left-hand side of that inequality. It can be found in this book, see Subsection 3.3.2 in Chapter 3.

Moreover, there is the reverse side of this approach. If one assumes that the integral

$$\int_{\mathbb{R}} \frac{|\widehat{g}(x)|}{|x|}\, dx$$

is finite, then g is the derivative of a function of bounded variation f which is absolutely continuous, $\lim_{|t|\to\infty} f(t) = 0$ and its Fourier transform is integrable.

This statement confirms the topic of the Fourier transform of a function of bounded variation to be sustainable and self-contained.

Introduction

Tools

We denote by $H^1(\mathbb{R})$ the real Hardy space. It can be defined in various ways, and we indeed shall need several, but what is of crucial importance for us is that here the Hilbert transform comes into play, since the real Hardy space $H^1(\mathbb{R})$ can be interpreted as
$$H^1(\mathbb{R}) := \{g \in L^1(\mathbb{R}) : \mathcal{H}g \in L^1(\mathbb{R})\},$$
where the Hilbert transform $\mathcal{H}g$ is defined as
$$\mathcal{H}g(x) := \frac{1}{\pi} \text{(P.V.)} \int_{\mathbb{R}} \frac{g(u)}{x-u} du$$
$$= \frac{1}{\pi} \lim_{\delta \downarrow 0} \int_\delta^\infty \{g(x-u) - g(x+u)\} \frac{du}{u}, \quad x \in \mathbb{R},$$
in the principle value sense. It is worth noting that if we assume that the derivative belongs to the Hardy space or certain of its subspaces (and is locally absolutely continuous), the assumption of bounded variation is redundant, since, on the one hand, belonging to one of these spaces provides $f' \in L^1$ and, on the other hand, integrability of the derivative along with the absolute continuity of the function makes it to be of bounded variation (at least, to be equal almost everywhere to such a function; see, e.g., [27, Ch. 1] or [92, 3.5.1]). We shall not discuss this issue in what follows and will not omit the assumption of bounded variation.

To illustrate the aforementioned advance, we will show what had happened with Theorem 2.8. The tools for a new approach is an operator we call a balance operator, having in mind that it allows to balance all the terms of the relations for the Fourier transform. It is defined by means of a generating function φ and takes on an appropriate function g the value
$$B_\varphi g(x) = \frac{1}{x^2} \int_0^\infty g\left(\frac{t}{x}\right) \varphi(t) dt.$$
This operator proved to be very convenient in many situations. For example, for $\varphi(t) = \sin t$, we have $\widehat{g}_s(x) = xB_sg(x)$ and, similarly, $xB_cg(x)$ is the cosine Fourier transform of g. Its only (obvious) property that we will need in this book is that for $g \in L^1(\mathbb{R}_+)$, we have $B_\varphi \in L^1(\mathbb{R}_+)$ provided
$$\int_0^\infty \frac{|\varphi(t)|}{t} dt < \infty.$$
Undoubtedly, certain substitutions and new notations may convert this operator to another one, more familiar, but it is convenient to use it in the present form under the new given name.

Sharp versions

In these terms, Theorem 2.8 reduces to Theorem 4.15 with exact formulas instead of relations with remainder terms:

Let $f \in BV_0[0, \infty) \cap LAC(0, \infty)$. Let almost everywhere

$$-\mathcal{H}_o(\mathcal{H}_e f')(t) = f'(t).$$

Then for the cosine Fourier transform of f, we have

$$\widehat{f_c}(x) = -B_s f'(x),$$

while for the sine Fourier transform an asymptotic formula holds: for $x > 0$,

$$\widehat{f_s}(x) = \frac{1}{x} f\left(\frac{\pi}{2x}\right) + B_s(\mathcal{H}_o f')(x) + B_S f'(x),$$

where B_S is generated by the function

$$S(t) = \frac{2}{\pi} \begin{cases} -t \int_0^\infty \frac{\sin s}{s(s+t)} ds, & 0 < t < \frac{\pi}{2}, \\ \int_0^\infty \frac{\sin s}{s+t} ds, & t \geq \frac{\pi}{2}. \end{cases}$$

The other relations, more recent than Theorem 2.8, also transform in the same manner, including a theorem for all functions of bounded variation with no relation to Hardy spaces.

Multivariate case

Of course, the picture would by no means be complete if the situation in the Euclidean space were not clarified, at least partially. This slip of the tongue/keyboard is not accidental, first of all because of the variety of the notions of multidimensional bounded variation. There is a variety of Hardy spaces as well. By the way, it seems that these two varieties exist not only because of the higher freedom in several dimensions but are closely tied to one another. We have chosen the Hardy variation as a universe in which the one-dimensional planet is located. There are a few reasons for this. Certainly, transparency is one of them. The other is that it leads to choosing a special product Hardy space as an extension of the one-dimensional Hardy space. This is not an optimal choice in many other topics but here it allows one to still use the Hilbert transform machinery. In Chapter 6, the above mentioned refinements of one-dimensional formulas open the way for precise multidimensional generalizations, which are preceded with old ones. Given partially because they seem to be different in different notation, the latter mainly recall how unpleasant these old extensions look with almost all terms being remainders. Now that newer versions really please the eye without being adjusted to specific notation but are sharp, they are equalities. We present meanwhile the continuation of the chosen chain. The next is a multidimensional generalization not only of Theorem 2.8 but also of the main result from Chapter 6:

Let $f : \mathbb{R}_+^n \to \mathbb{C}$ be of bounded Hardy variation on $\mathbb{R}_+^n$ and let f vanish at infinity along with all $D^\eta f$ except $\eta = \mathbf{1}$. Let also f and the same $D^\eta f$ be locally absolutely continuous. In addition, let for all the derivatives the inverse formula for the Hilbert transform holds almost everywhere in each variable. Then

$$(-1)^{|\eta|}\widehat{f}_\eta(x) = \left(\prod_{j:\eta_j=0}\frac{1}{x_j}\right) B_s^\eta D^\eta f\left(x_\eta, \left(\frac{\pi}{2x}\right)_{\mathbf{1}-\eta}\right)$$

$$+ \sum_{\substack{\chi:\chi_i=0 \ if \ \eta_i=1, \\ \chi\neq \mathbf{0},\mathbf{1}}} \left(\prod_{j:\chi_j=1}\frac{1}{x_j}\right)$$

$$\times \prod_{j:(\mathbf{1}-\eta-\chi)_j=1} \left(B_s^j \mathcal{H}_o^j \frac{\partial}{\partial x_j} + B_S^j \frac{\partial}{\partial x_j}\right)$$

$$\times B_s^\eta D^\eta f\left(x_{\mathbf{1}-\chi}, \left(\frac{\pi}{2x}\right)_\chi\right)$$

$$+ \prod_{j:(\mathbf{1}-\eta)_j=1} \left(B_s^j \mathcal{H}_o^j \frac{\partial}{\partial x_j} + B_S^j \frac{\partial}{\partial x_j}\right) B_s^\eta D^\eta f(x).$$

Picture and details

The three complete formulations given above and the relations between them present a comprehensive idea of our intentions and approaches to the various results on the Fourier transform of a function of bounded variation. Nevertheless, before inviting the reader to the main body of this book, we wish to explain briefly a couple of different structural details.

Much place is given to the background, both in dimension one and in several dimensions. There are many reasons for this. One of them is that we wish these parts not only to cover all the needed machinery but to be instructive. We sometimes give details that may seem unnecessary for professionals, first of all in the starting chapters of each part, where relevant background is presented; however, we wish certain parts of the text to be also appropriate for special courses for (even) undergraduate students and their self-study. While the basics on functions of bounded variation may by no means be surprising, more or less repeating well known facts and are given only to make the presentation self-contained, the collection of facts on the Hardy space and Hilbert transform does not add much to the known information but may be somewhat surprising even for the specialists. Not because of the deepness of the results but mostly because some of them have "accurately" been side-stepped in the literature. When preparing this section, what was most surprising was the need to prove them rather than just to refer to them. In the corresponding opening Chapter 5 in Part II, special attention is paid to notation. This is clear, since proper notation is one of the keystones of successive work in several dimensions. In particular, we hope that the indicator vectors introduced as the groundwork of our formulations are up to this task.

As mentioned, most of the book concerns locally absolutely continuous functions. However, there is a strong incentive to make the presentation even richer by removing the condition of absolute continuity. It seems rather obvious that this can be done by replacing integrability by parts in the Riemann integral that many of the proofs start with by the Riemann-Stieltjes integrability. However, there is a pitfall when the integrability of the Hilbert transform of the measure df comes into play. This leads to nowhere, since such functions are necessarily absolutely continuous, which is akin to the theorems of the brothers Riesz. Therefore, the world "below" $H^1(\mathbb{R})$ remains "untouched" in the sense of the results of Wiener and Wintner, Ivashev-Musatov, and others.

The next to last chapter returns us to the derivation of the whole business, the theory of integrability of trigonometric series. We explain how such problems can be reduced to those for the Fourier transforms of functions of bounded variation. We prove the known one-dimensional "bridge" between the two theories and a fairly recent analog for functions with bounded Hardy's variation. Each result is illustrated with an application to trigonometric series, very recent in dimension one and more or less known for several dimensions.

The final chapter of the book stands somewhat apart of the mainstream of the text. However, it would be strange to ignore the existence of such a transition between the one-dimensional case and the multivariate one as radial functions. They usually possess the properties of both cases. Note that before giving really radial extensions of the obtained one-dimensional results, we present a general necessary condition for the Fourier transform to be integrable. It deals with the radial part of a function and uses the T-transform mentioned not even once throughout.

In conclusion, the book presents a comprehensive picture of the behavior of the Fourier transform of a function of one variable with bounded variation and its natural extensions to functions of several variables. It is known that such results have numerous applications in approximation, summability and other areas of analysis. Most of the results are very recent and have either been published or accepted for publication in various journals. The list of such recent publications more or less reflects the contents of this book: [136], [128], [115], [118], [197], [120], [121], [122], [123], [124], [125], [126], [127], [133].

In his review on "Classical and multilinear harmonic analysis" by C. Muscalu and W. Schlag, C. Demeter wrote in the Bull. Amer. Math. Soc. **52** (2015), 159–165, that "If the "Adam" of the field (harmonic analysis) was the idea of Fourier decomposition, a good candidate for the "Eve" is the Hilbert transform... The two volumes of the book under review are largely devoted to exploring the offspring of their marriage." In these terms one of the main trends of this book may apparently be characterized as the search for (the traces of) Lilith or the serpent. Has something like this been found? "There's some debate about that", as Lewis Carroll's Alice said once. But the process of looking for less usual relations between and peculiarities of the two transforms was intriguingly interesting for the author, who hopes that to read about this will be interesting to the readers. Let us proceed to the stories and details.

Part I:
One-dimensional Case

This part will be purely one-dimensional. In fact, the second part in which multidimensional generalizations will be given cannot exist, even in presentation, without this one-dimensional matter. Most of the ideas crucial for the given results are in dimension one, though their importance will be revealed in the multivariate setting.

Chapter 1

A toolkit

In this chapter we present the main ingredients for our work, and the machinery we are going to apply. Some parts of it are very brief while some other are very detailed, especially those concerned with the Hilbert transform and Hardy spaces. The point is that these parts contain facts and results that have not appeared together in one source. Most of them are known but apparently do not play the same role they do in our study. Some others are not to be found in the literature at all. Such are, for instance, the example of an odd integrable function with non-integrable Hilbert transform or the example of a function not belonging to the real Hardy space but with finite left-hand side in the Fourier–Hardy inequality (see Subsection 3.3.2 in Chapter 3). In any case, we hope that this collection will be of interest and of certain benefit in its own right.

1.1 Functions of bounded variation

In this section, we collect information on functions of bounded variation. This notion goes back to Jordan, has underwent a change in its popularity and use, mainly because measures rather than functions are better in many situations, but still plays a notable role in a variety of topics. One of the ways to motivate our interest in this class of functions is as follows. There is no doubt about how important monotone functions are. However, the class of such functions is not closed with respect to the simplest algebraic operations. The class of functions of bounded variation is a proper extension of monotone functions. We also take this opportunity to mention recent attempts to generalize monotonicity in different ways for various problems; see the survey type paper [137] and recent results of that kind in such a classical area as number series in [138].

We start with classical facts for functions on a finite interval $[a, b]$. Since a concentrated version of basics is needed, we follow some of the sources where such a release is given, see, e.g., [151], [142]. As mentioned, even a whole book on this subject has recently appeared [9]. Of course, it is recommended for the study of the subject but to have the basics concentrated another sources might be more convenient.

Consider an arbitrary partition τ of $[a,b]$ by means of the points $t_0 = a < t_1 < \cdots < t_m = b$ and, for a function f defined on $[a,b]$, set

$$S_\tau = \sum_{k=0}^{m-1} |f(t_{k+1}) - f(t_k)|.$$

Definition 1.1. The value $\sup_\tau S_\tau$ is called the *total variation* of the function f on the interval $[a,b]$ and is denoted by $Vf := V_{[a,b]}f$. If $V_{[a,b]}f$ is finite, an alternative notation is $f \in BV([a,b])$ (or another set in place of $[a,b]$), f is called a *function of bounded variation*.

It is clear that by adding new partition points, S_τ may only increase. Let us list additional properties of the total variation and of functions with bounded variation.

(1) $V_{[a,b]}f \geq |f(b) - f(a)|$.

(2) A monotone function f is of bounded variation, with

$$V_{[a,b]}f = |f(b) - f(a)|.$$

(3) A linear combination of functions of bounded variation is again a function of bounded variation, with

$$V_{[a,b]}(f+g) \leq V_{[a,b]}f + V_{[a,b]}g$$

and

$$V_{[a,b]}(\alpha f) = |\alpha| V_{[a,b]}f, \quad \alpha \in \mathbb{R}.$$

(4) The product of two functions of bounded variation is again a function of bounded variation; the quotient of two functions of bounded variation is again a function of bounded variation provided that the denominator is bounded away from zero.

A less obvious property of the total variation is additivity.

(5) If $a < c < b$, then $V_{[a,b]}f = V_{[a,c]}f + V_{[c,b]}f$.

It applies both to the case of bounded and unbounded variation.

As one can see from (2) and (3), the difference of two increasing functions is a function of bounded variation. It follows from the last property that the converse is also true.

1.1. Functions of bounded variation

(6) A function is of bounded variation if and only if it can be written as the difference of increasing functions (a particular case of Jordan decomposition for measures). In addition, a function of bounded variation can be written as the difference of increasing functions that are continuous at the same points as f is. All these are for real-valued functions; for a complex-valued function these assertions hold for the real and imaginary part.

This property implies, from the known properties of monotone functions, that, in particular, the set of points of discontinuity of a function of bounded variation is at most countable.

(7) Each function of bounded variation is bounded.

The notion of absolute continuity is in order.

Definition 1.2. A function f defined on an interval $[a,b]$ is called absolutely continuous on it if for every positive number ε, there is a positive number δ such that whenever a finite sequence of pairwise disjoint sub-intervals (x_k, y_k) of $[a,b]$ with $x_k, y_k \in [a,b]$ satisfies

$$\sum_k (y_k - x_k) < \delta,$$

then

$$\sum_k |f(y_k) - f(x_k)| < \varepsilon.$$

The collection of all absolutely continuous functions on $[a,b]$ is denoted by $AC([a,b])$. There is an equivalent definition of absolute continuity, of which we shall mostly make use. If one of them is chosen to be the definition, the other can be derived from it as a property of an absolutely continuous function.

Definition 1.3. A function f defined on an interval $[a,b]$ is called absolutely continuous on it if it can be written in the form

$$f(x) = F(c) + \int_c^x h(t)\,dt, \quad x \in [a,b],$$

where $c \in [a,b]$ and h is Lebesgue integrable on $[a,b]$.

The notion of absolute continuity can be defined in a similar way in the case where the interval is not closed; however, the function h may be not integrable on it but only *locally* integrable, that is, integrable on every *closed subinterval*.

The following Lebesgue theorem gives the crucial property of absolutely continuous functions.

Theorem 1.4. *If f is a function that is absolutely continuous on an interval $[a,b]$, then it is differentiable almost everywhere, its derivative is (locally) integrable, and*

$$f(y) - f(x) = \int_x^y f'(t)\,dt$$

for any $x, y \in [a,b]$.

To compare with the more general case, it is worth mentioning that if we know only that f is an increasing function, then its derivative is measurable and

$$\int_x^y f'(t)\, dt \leq f(y) - f(x),$$

which gives the integrability of the derivative (see, e.g., [151, Th.5, Ch.VIII, §2]).

Connections between functions of bounded variation and absolutely continuous functions have been begun to be studied long ago, see, e.g., [159]. Today these connections are known in full. First, absolutely continuous functions and functions of bounded variation are related in the following way.

Theorem 1.5. *If a function f is absolutely continuous on $[a,b]$, then it is of bounded variation, with*

$$V_{[a,b]}f = \int_a^b |f'(t)|\, dt.$$

Note that a function f absolutely continuous on $[a,b]$ can be written as the difference of absolutely continuous increasing functions.

We now continue to list the properties of functions of bounded variation. The structure of a function of bounded variation is revealed by Lebesgue's decomposition.

(8) Every function of bounded variation f can be represented as

$$f(t) = f_{jump}(t) + f_{AC}(t) + f_{sing}(t), \tag{1.6}$$

where f_{jump} is the function of the jumps of f, f_{AC} is an absolutely continuous function and f_{sing} is a *singular function*. The latter means that f_{sing} is a non-constant function and its derivative vanishes almost everywhere. Of course, if f is a continuous function, f_{jump} does not exist in (1.6) as well as both f_{jump} and f_{sing} are absent if f is absolutely continuous.

Let us now make a few remarks on functions of bounded variation on infinite intervals. Good sources where certain peculiarities of this instance are studied are [167] or [36]. The total variations $V_{[a,\infty]}f$ and $V_{[-\infty,b]}f$ can be defined by

$$V_{[a,\infty)}f = \lim_{b\to\infty} V_{[a,b]}f$$

and

$$V_{(-\infty,b]}f = \lim_{a\to-\infty} V_{[a,b]}f.$$

We mostly deal with functions f of bounded variation on $\mathbb{R}_+ = [0,\infty)$, and vanishing at infinity, $\lim_{t\to\infty} f(t) = 0$, written $f \in BV_0[0,\infty)$, and locally absolutely continuous on $(0,\infty)$, written $f \in LAC(0,\infty)$. We will denote these by

$$f \in BV_0[0,\infty) \cap LAC(0,\infty).$$

1.1. Functions of bounded variation

We shall use the standard notation

$$\|f\|_{BV} = \int_{\mathbb{R}} |df(t)|. \tag{1.7}$$

Contrary to the notation $BV_0(\mathbb{R})$, that is, the space of BV functions vanishing at both $+\infty$ and $-\infty$, we will denote the space of BV functions vanishing at $-\infty$ regardless of their behavior at $+\infty$, by BV_-, though the norm will be kept as $\|\cdot\|_{BV}$. In fact, BV_0 is the set of BV_- functions for which

$$\int_{\mathbb{R}} df(t) = 0.$$

For $g \in L^1$, the function f defined by

$$f(t) = \int_{-\infty}^{t} g(s)\, ds$$

is an element of BV_-, and thus $AC(\mathbb{R}) \subset BV_-$.

This also allows one to proceed to integration by parts with respect to a measure generated by a function of bounded variation, the so-called Stieltjes integration. As in the "usual" case, there is the Riemann–Stieltjes integration and more general Lebesgue–Stieltjes integration.

The formula for the Lebesgue–Stieltjes integration by parts reads as follows:

$$\int_{a+}^{a+b} f(x)\, dg(x) = (fg)(a+b) - (fg)(a) - \int_{a+}^{a+b} g(x-)\, df(x). \tag{1.8}$$

Of course, $g(x-)$ denotes the left limit of g at x. Standard assumptions on f and g might both be of bounded variation on every finite interval of $\mathbb{R}$ (see, e.g., [87, 21.67 and 21.68]). Of course, under an additional assumption, say, absolute continuity, there might be appropriate changes in this formula.

Much interesting information can be found in Chapter VIII of [151] in the Riemann–Stieltjes setting. For example, the integral

$$\int_a^b f(t)\, dg(t) \tag{1.9}$$

exists if f is continuous and g is of bounded variation. If both functions are only of bounded variation, one should be careful if there is a point of discontinuity common for both functions; otherwise no problems appear. It is also worth mentioning an interesting sufficient condition for the existence of (1.9) due to Kondurar and given as an exercise to the same chapter: one of the functions satisfies the Lipschitz condition of order α, while the other of order β, and $\alpha + \beta > 1$.

Remark 1.10. In general, Lebesgue–Stieltjes integrals are usually considered with respect to arbitrary Borel measures rather than for functions of bounded variation such as in this work. But there is a close connection between Borel measures and BV-classes. Indeed, apart from a different normalization in the interior points (instead of

$$f(t) = \frac{f(t+0) + f(t-0)}{2}$$

one assumes f to be left-continuous, i.e., $f(t) = f(t-0)$), there is a one-to-one correspondence between bounded Borel measures on $\mathbb{R}$ and functions $f \in BV_-$ and between Borel measures on $[a, b]$ and functions of bounded variation on $[a, b]$ with $f(a) = 0$.

As the famous Russian bard Vladimir Vysotsky sang in one of his songs: "I do have a measure, even two..." Of course, in a different connection and not about mathematical measures, but it sounds quite appropriate.

Similar correspondence exists in the multivariate case too, but we shall not touch on this issue in the second part.

1.2 Fourier transform

We cannot avoid giving certain basics on the Fourier transform. There apparently will not be any novelties or surprises here but it is meaningful to aggregate corresponding machinery about the Fourier transforms we are going to use. In any case, only certain specific and compulsory facts will be given in dimension one. More general and regular data will be presented in Chapter 5 in the multivariate setting.

Let

$$\widehat{g}(x) = \int_{-\infty}^{\infty} g(t) e^{-ixt} dt \qquad (1.11)$$

be the Fourier transform of g; let it exist in that or another sense, no need for greater accuracy at this point. One should observe here that we define it without any coefficient before the integral, which leads to $\frac{1}{2\pi}$ before the integral in the inverse Fourier transform

$$\check{h}(t) = \frac{1}{2\pi} \int_{-\infty}^{\infty} h(t) e^{ixt} dx. \qquad (1.12)$$

One of the main questions in Fourier Analysis is whether and in what sense (1.12) restores $g(t)$ if $h(x) = \widehat{g}(x)$. Or, for which functions g this is true.

It is reasonable to ask at this point how slow $\widehat{g}(x)$ goes to 0 as $|x| \to \infty$ provided $g \in L^1(\mathbb{R})$ (and it does go, this fact is known as the Riemann–Lebesgue lemma). The answer is as one pleases! It is known that an arbitrary rate of convergence to 0 is possible. Indeed, a classical result due to Pólya (see, e.g., [141] or [198]) says that each even, bounded and convex function on $[0, \infty)$ monotone

1.2. Fourier transform

decreasing to zero is the Fourier transform of an integrable function. The proof will be presented later on in Chapter 2 as a consequence of general results. More sophisticated examples of that kind can be found in [128]. In other words, nothing but uniform continuity and the Riemann–Lebesgue lemma can be said on the Fourier transform of an integrable function. Again, for more details, see Chapter 5.

For completeness, let us outline how the Riemann–Lebesgue lemma can be proved. It is immediate for the indicator function of an interval and, similarly, for the indicator function of a finite union of intervals. Then, approximating the given function f by a finite linear combination of the indicators of intervals, that is, by step functions, in the L^1 norm, we let the Fourier transform $\widehat{f}$ become small because of the error of approximation and smallness of the Fourier transform of the approximant.

We shall mainly study the cosine Fourier transform

$$\widehat{f}_c(x) = \int_0^\infty f(t) \cos xt \, dt \qquad (1.13)$$

and the sine Fourier transform

$$\widehat{f}_s(x) = \int_0^\infty f(t) \sin xt \, dt, \qquad (1.14)$$

and their integrability properties rather than the general Fourier transform (1.11) that will appear only from time to time, with results that mostly are consequences of the obtained results for (1.13) and (1.14). Our eternal question will be whether the Fourier transform is integrable or not. In fact, a similar question can be asked:

Given f uniformly continuous and vanishing at infinity, whether is it representable as the Fourier integral of an integrable function g, written

$$f(t) = \int_{\mathbb{R}} g(x) e^{itx} dx? \qquad (1.15)$$

It is said in this case that f belongs to the Wiener space (algebra) $W_0(\mathbb{R})$. Of course, in many situations g can be understood, in that or another sense, as the Fourier transform of f and the last formula as the Fourier inversion. A comprehensive overview of these problems is given in [128]; it is worth noting that this name and notation is frequently used for a different algebra, see, e.g., [162], [54] or [77].

To get a flavor of functions with integrable Fourier transform (or of those belonging to W_0), we note that such a function necessarily possesses a certain smoothness. If, say, the cosine Fourier transform $\widehat{f}_c$ is integrable on $\mathbb{R}_+$, then the integrals

$$\int_\delta^{\frac{x}{2}} \frac{f(x+t) - f(x-t)}{t} dt \qquad (1.16)$$

are uniformly bounded (for the well-known prototype for Fourier series, see [97, Ch.II, §10], while in [198, 3.5.5] an even more subtle result of this type is given). Indeed, expressing $f(x+t)$ and $f(x-t)$ via the Fourier inversion, we obtain

$$\left| \int_\delta^{\frac{x}{2}} \frac{f(x+t) - f(x-t)}{t} dt \right|$$

$$= \frac{1}{\pi} \left| \int_\delta^{\frac{x}{2}} \frac{1}{t} \int_0^\infty \widehat{f}_c(u)[\cos u(x+t) - \cos u(x-t)] \, du \, dt \right|$$

$$\leq \frac{2}{\pi} \int_0^\infty |\widehat{f}_c(u)| \left| \int_\delta^{\frac{x}{2}} \frac{\sin t}{t} dt \right| du.$$

The last integral on the right is uniformly bounded. The result and proof for the sine and general Fourier transform are the same.

Of course, it is worth mentioning that the inverse formula is true almost everywhere if the Fourier transform is integrable, which is not the case for the Fourier transforms of an integrable function in general. The latter is illustrated in the most impressive way by Kolmogorov's famous example in [103]. It is for the Fourier series but it is the same for the Fourier transform and delivers an integrable function whose inverse integral diverges EVERYWHERE.

For functions $f \in L^p(\mathbb{R})$, with $1 < p < 2$, we will use the Hausdorff–Young inequality. It reads, with $\frac{1}{p'} = 1 - \frac{1}{p}$, as

$$\|\widehat{f}\|_{L^{p'}(\mathbb{R})} \leq C_p \|f\|_{L^p(\mathbb{R})}. \tag{1.17}$$

We shall apply the Hausdorff–Young inequality in this form, but with the constant 1 rather than C_p on the right, which is not sharp. A sharp constant C_p is due to Beckner [17] (an earlier partial result is due to K. Babenko [11]), but it is of no importance for our estimates. Nevertheless, recall that the Babenko–Beckner constant is

$$C_p = \frac{p^{\frac{1}{2p}}}{(p')^{\frac{1}{2p'}}}.$$

We note that the Hausdorff–Young inequality is one of the possible avenues where interpolation theory comes into play. More precisely, it can be proved by interpolating between the obvious $L^1 - L^\infty$ estimate for the Fourier transform and the $L^2 - L^2$ Plancherel one.

The next assertion not only relates the main ingredients of our approach but also justifies the general interest in the study of the Fourier transform just for functions of bounded variation.

Proposition 1.18. *Let*

$$\int_\mathbb{R} \frac{|\widehat{g}(x)|}{|x|} dx < \infty. \tag{1.19}$$

Then g is the derivative of a function of bounded variation f which is absolutely continuous, $\lim\limits_{|t|\to\infty} f(t) = 0$, and the Fourier transform of g is integrable.

Proof. Denoting
$$f(t) = \int_{-\infty}^{t} g(u)\,du,$$
we integrate by parts as follows
$$\int_{-\infty}^{\infty} g(t)e^{-ixt}\,dt = \left[e^{-ixt}\int_{-\infty}^{t} g(u)\,du\right]_{-\infty}^{\infty}$$
$$+ ix\int_{-\infty}^{\infty}\left[\int_{-\infty}^{t} g(u)\,du\right]e^{-ixt}\,dt = ix\int_{-\infty}^{\infty} f(t)e^{-ixt}\,dt.$$

The integrated terms vanish since, by (1.19), g must have mean zero (cf. 1.49). It follows now from (1.19) that the L^1 norm of $\widehat{f}$ is finite. □

This proposition is also a kind of preamble to the future study of Hardy's inequality (1.61).

Of course, there are many other reasons for the Fourier transform of a function of bounded variation to be studied thoroughly. For example, the Fourier multiplier is one of the central notions in analysis. It is said that a (bounded or L^∞) function is a $X \to Y$ Fourier multiplier if the operator defined by means of the relation
$$\widehat{M_m f}(x) = m(x)\widehat{f}(x)$$
is bounded taking X to Y. One of the important facts in the theory of multipliers is that if m is a function of bounded variation on $\mathbb{R}$, then m is a Fourier multiplier on L^p (preserves L^p) for $1 < p < \infty$ (see, e.g., [50, Corollary 3.8]).

Last but not least, the Fourier transform of a function of bounded variation is well-defined; to see this, it suffices to integrate by parts in the Stieltjes sense. Indeed, for $f \in BV_0(\mathbb{R})$, we have
$$\widehat{f}(x) = \frac{1}{ix}\int_{\mathbb{R}} e^{-ixt}\,df(t), \tag{1.20}$$

and the right-hand side is well-defined because of (1.7). To distinguish (1.20) from the "standard" transform, (1.20) is often called the Fourier–Stieltjes transform.

1.3 Hilbert transform

The Hilbert transform of a function g is
$$\mathcal{H}g(x) = \frac{1}{\pi}\int_{\mathbb{R}} \frac{g(t)}{x-t}\,dt, \tag{1.21}$$

where the integral is understood in the improper (principal value) sense, as

$$\mathcal{H}g(x) = \lim_{\delta \to 0+} \mathcal{H}_\delta(x),$$

with

$$\mathcal{H}_\delta g(x) = \frac{1}{\pi} \int_{|t-x|>\delta} \frac{g(t)}{x-t} \, dt$$

being the truncated Hilbert transform. Note that the study of the maximal Hilbert transform

$$\mathcal{H}^* g(x) = \sup_{\delta > 0} |\mathcal{H}_\delta g(x)|$$

is very important in many problems. However, it will not be dealt with here.

Pandey begins his book [157] with a list of fields where the Hilbert transform arises. It is worth repeating here:

i. Signal processing (the Hilbert transform of periodic functions)

ii. Metallurgy (Griffith crack problem and the theory of elasticity)

iii. Dirichlet boundary value problems (potential theory)

iv. Dispersion relation in high energy physics, spectroscopy, and wave equations

v. Wing theory

vi. The Hilbert problem

vii. Harmonic analysis

In the sequel, he explains some of these applications in more detail. However, we shall deal with the Hilbert transform within the framework of harmonic analysis only. Along with the Fourier transform, the Hilbert transform is the most important operator in harmonic analysis. There is an immense amount of motivations, or, in other words, problems that motivate the use of these tools. One of them, of course, comes from complex analysis, where the Hilbert transform characterizes analytic functions in the upper half-plane. In addition to the above, recall also that it appears in the inversion formula of the Radon transform, see, e.g., [152, Ch. II]. We need not even mention the role of computer tomography in our life today.

We shall denote the way of integration by means of which the Hilbert transform is defined by (P.V.) before the integral sign, if needed. Let us consider a few simple examples. Each of them will be used in the sequel.

Example 1.22. The Hilbert transform of a constant function is identically zero. This is because of the symmetry in the principal value approach.

Example 1.23. The Hilbert transform of the indicator function $\chi_{[a,b]}$ exists except at the two points a and b and is given by

$$\mathcal{H}\chi_{[a,b]}(x) = \frac{1}{\pi} \ln \frac{|x-a|}{|x-b|}. \tag{1.24}$$

1.3. Hilbert transform

Let $x \neq a, b$ and $\delta < \min\{|x-a|, |x-b|\}$. If $x - b > 0$ or $x - a < 0$, then (1.24) follows immediately by simple integration. If $a < x < b$, then

$$\mathcal{H}\chi_{[a,b]}(x) = \frac{1}{\pi} \lim_{\delta \to 0+} \left(\ln \frac{|x-a|}{\delta} + \ln \frac{\delta}{|x-b|} \right),$$

which implies (1.24). In this last case, we essentially use that the transform is understood in the principal value sense.

Example 1.25. The Hilbert transform of $g(t) = \sin at$, $a > 0$, on the real line is given by $\mathcal{H}g(x) = -\cos ax$.

We have

$$\lim_{\delta \to 0+} \frac{1}{\pi} \int_{|x-t| \geq \delta} \frac{\sin at}{x-t} \, dt = \frac{1}{\pi} \lim_{\delta \to 0+} \int_{|t| \geq \delta} \frac{\sin(ax - at)}{t} \, dt$$

$$= \frac{\sin ax}{\pi} \lim_{\delta \to 0+} \int_{|t| \geq \delta} \frac{\cos at}{t} \, dt - \frac{\cos ax}{\pi} \lim_{\delta \to 0+} \int_{|t| \geq \delta} \frac{\sin at}{t} \, dt.$$

The first integral on the right-hand side vanishes because of the oddness of the integrand. The second one is well known and equals π. Along with the previous factor, it gives the required result.

Example 1.26. In a similar way, the Hilbert transform of $g(t) = \cos at$, $a > 0$, on the real line is given by $\mathcal{H}g(x) = \sin ax$.

In these examples, we were able to use the definition of the Hilbert transform. However, we must have a deeper knowledge of the existence of the Hilbert transform for functions of various classes. Prior to this, we shall try to realize how the Hilbert transform may appear in general. What we mean is that there are various ways in which the Hilbert transform comes into play. Not that one of them is definitely better than the others – it is much a question of the topic and taste. We decided to restrict ourselves to one of the oldest and somewhat "naive" approaches in regard to rigorousness. It is one of the possible heuristic approaches and it is what the next subsection is about.

1.3.1 Fourier transform weakly generates Hilbert transform

We present a natural formal way how the Hilbert transform appears in close connection with the Fourier transform, following Chapter V of Titchmarsh's celebrated book [186]. Of course, there are different ways to introduce the Hilbert transform; they are described in the very beginning of [100]. However, we chose the one that is one of the nearest to the relations between the two transforms. On the other hand, it does not avoid completely the complex analysis approach.

The Fourier integral formula can be written as an analog of the Fourier series in the following manner (Fourier inversion):

$$g(t) = \int_0^\infty [a(x) \cos tx + b(x) \sin tx] \, dx, \qquad (1.27)$$

where
$$a(x) = \frac{1}{\pi} \int_{-\infty}^{\infty} g(t) \cos xt \, dt$$
and
$$b(x) = \frac{1}{\pi} \int_{-\infty}^{\infty} g(t) \sin xt \, dt.$$

Formally, the integral in (1.27) is the limit, as $u \to 0$, of the integral
$$\int_0^{\infty} [a(x) \cos tx + b(x) \sin tx] e^{-ux} \, dx = U(t, u),$$
while the latter is the real part of the function
$$\int_0^{\infty} [a(x) - ib(x)] e^{izx} \, dx = \Phi(z),$$
with $z = t + iu$. The imaginary part of the function $\Phi(z)$ is
$$-\int_0^{\infty} [b(x) \cos tx - a(x) \sin tx] e^{-ux} \, dx = V(t, u).$$

Denoting $V(t, 0) = \mathcal{H}g(t)$, we obtain
$$\mathcal{H}g(t) = -\int_0^{\infty} [b(x) \cos tx - a(x) \sin tx] \, dx$$
$$= \frac{1}{\pi} \int_0^{\infty} \int_{-\infty}^{\infty} \sin(t - v)x \, g(v) \, dv \, dx. \qquad (1.28)$$

The integral on the right-hand side of (1.28) is called the conjugate integral of the Fourier integral. It can formally be derived from (1.27) by replacing a and b with $-b$ and a, respectively.

It can also be formally established that
$$\widehat{\mathcal{H}g}(-x) = i \operatorname{sign} x \, \widehat{g}(x). \qquad (1.29)$$

The latter gives rise to a possible different way to define the Hilbert transform via the Fourier multipliers theory, an object we have already faced, at least on the level of definitions and basics. More precisely, for an appropriate class, the Hilbert transform can be defined by means of (1.29), with the multiplier $i\operatorname{sign}(\cdot)$ (see, e.g., [158], which might be instructive for the previous section too, as well as for some other topics).

Further, and once again formally,
$$\mathcal{H}g(t) = \frac{1}{\pi} \int_0^{\infty} \frac{g(t-v) - g(t+v)}{v} \, dv = \frac{1}{\pi} \int_{-\infty}^{\infty} \frac{g(v)}{v - t} \, dv. \qquad (1.30)$$

1.3. Hilbert transform

In parallel,

$$g(t) = -\frac{1}{\pi}\int_0^\infty \frac{\mathcal{H}g(t-v) - \mathcal{H}g(t+v)}{v}\,dv = -\frac{1}{\pi}\int_{-\infty}^\infty \frac{\mathcal{H}g(v)}{v-t}\,dv. \qquad (1.31)$$

This duality was first noticed by Hilbert, hence the pair of these transforms are called the Hilbert transforms. What is given above is, in general, much of the formal theory, at least initial, of the Hilbert transform. However, the change of the order of such operations as limit or integration has never been justified during that presentation. Of course, no universal way exists for this, all depends upon the setting in which the game takes place. There are many of them which proved to be useful. Justifying the above operations in an appropriate setting is, in a sense, the corresponding "genuine" theory. One more thing that may vary from setting to setting and is sometimes a matter of taste is the sign before the integral and sometimes i or its absence before the integral.

The Hilbert transform (and, correspondingly, the above formulas) is well-defined and exists almost everywhere for various classes of functions, say, for L^p, with $1 \le p < \infty$. There are numerous sources where this fact is proven in that or another way (see, e.g., [186] or [144]). We prefer to follow the way how it is given in [36, 8.1.1], where this assertion is proven for the more general Hilbert transform

$$\mathcal{H}df(x) = \frac{1}{\pi}\int_{\mathbb{R}} \frac{df(t)}{x-t}, \qquad (1.32)$$

the Hilbert transform of the measure df generated by a function of bounded variation f subject to the condition $\lim_{t\to-\infty} f(t) = 0$. In the further theorems on the behavior of the Fourier transform, its counterpart $\lim_{t\to+\infty} f(t) = 0$ will be used. Let us mention that, for example, that (1.32) was widely used as early as in 1922 by R. Nevanlinna in [153].

1.3.2 Existence almost everywhere

First of all, to distinguish this case from the previous one, the transform in (1.32) is often called the *Hilbert–Stieltjes transform*. It also makes us recall that integration in (1.32) is, in fact, Stieltjes integration. For the following results, see [140] or [36].

Theorem 1.33. *The Hilbert–Stieltjes transform* (1.32) *exists almost everywhere. Moreover, for every $M > 0$,*

$$\operatorname{meas}\{x : |\mathcal{H}df(x)| > M\} \le \frac{128}{M}\|f\|_{BV}. \qquad (1.34)$$

To prove this theorem, we need the following lemmas.

Lemma 1.35. *Let $d_k > 0$, $a_k \in \mathbb{R}$, $1 \le k \le m$, and*

$$g(x) = \sum_{k=1}^{m} \frac{d_k}{x - a_k}. \tag{1.36}$$

Then

$$\operatorname{meas}\{x : |g(x)| > M > 0\} = \frac{2}{M} \sum_{k=1}^{m} d_k, \tag{1.37}$$

and this set consists precisely of $2m$ intervals.

Proof. Since $g(a_k-) = -\infty$, $g(a_k+) = +\infty$ and $g'(x) < 0$ for all $x \ne a_k$, $1 \le k \le m$, there are precisely m points b_k such that $g(b_k) = M$ with $a_k < b_k < a_{k+1}$, $1 \le k \le m-1$, and $a_m < b_m$. The set where $g(x) > M$ therefore consists of the intervals (a_k, b_k) and has total length

$$\sum_{k=1}^{m}(b_k - a_k) = \sum_{k=1}^{m} b_k - \sum_{k=1}^{m} a_k. \tag{1.38}$$

But the numbers b_k are the roots of the equation

$$\sum_{k=1}^{m} \frac{d_k}{x - a_k} = M;$$

and hence multiplying by $\prod_{k=1}^{m}(x - a_k)$, we have

$$\sum_{k=1}^{m} d_k \left[\prod_{j \ne k}(x - a_j) \right] = M \prod_{k=1}^{m}(x - a_k)$$

or

$$Mx^m - \left[M\sum_{k=1}^{m} a_k + \sum_{k=1}^{m} d_k \right] x^{m-1} + \cdots = 0.$$

If we compare the coefficients of the last equation with those of $\prod_{k=1}^{m}(x - b_k) = 0$, then

$$\sum_{k=1}^{m} b_k = \sum_{k=1}^{m} a_k \frac{1}{M} \sum_{k=1}^{m} d_k.$$

Substituting into (1.38) and combining this with the corresponding result for the set where $g(x) < -M$, we see that (1.37) follows. □

1.3. Hilbert transform

Lemma 1.39. *Let $f \in BV_-$ be real-valued. If $(x_k - \delta_k, x_k + \delta_k)$, $1 \leq k \leq m$, are disjoint intervals such that*

$$\left| \left(\int_{-\infty}^{x_k - \delta_k} + \int_{x_k+\delta_k}^{\infty} \right) \frac{df(t)}{x_k - t} \right| > M > 0, \tag{1.40}$$

then

$$\sum_{k=1}^{m} \delta_k \leq \frac{16}{M} \|f\|_{BV}.$$

Proof. First we observe that the integrals (1.40) are well-defined and bounded by

$$\frac{1}{\delta_k} \|f\|_{BV}, \quad 1 \leq k \leq m.$$

Hence they may be approximated arbitrarily closely by finite Riemann-Stieltjes sums, if the norm of the subdivision is sufficiently small. Thus let u_j, $1 \leq j \leq N$, be a finite subdivision including the points $x_k - \delta_k$, x_k, $x_k + \delta_k$, $1 \leq k \leq m$, such that, according to (1.40), for each k, $1 \leq k \leq m$,

$$\left| \sum_{j \notin I_k} \frac{f(u_{j+1}) - f(u_j)}{y - u_j} \right| > M > 0 \tag{1.41}$$

holds for $y = x_k$, where the set I_k of indices omitted is defined by

$$\bigcup_{j \in I_k} (u_j, u_{j+1}) = (x_k - \delta_k, x_k + \delta_k).$$

Now, suppose that f is monotone increasing. Then the left-hand side of (1.41) is a decreasing function of y for $x_k - \delta_k < y < x_k + \delta_k$. Hence, if the sum is positive (negative), one of the following inequalities is therefore satisfied:

$$\left| \sum_{j=1}^{N-1} \frac{f(u_{j+1}) - f(u_j)}{y - u_j} \right| > \frac{M}{2}$$

or

$$\left| \sum_{j \in I_k} \frac{f(u_{j+1}) - f(u_j)}{y - u_j} \right| > \frac{M}{2}.$$

Since $f(u_{j+1}) - f(u_j) \geq 0$, we may apply Lemma 1.35, and summing over k it follows that

$$\sum_{k=1}^{m} \delta_k \leq \frac{4}{M} \sum_{j=1}^{N-1} [f(u_{j+1}) - f(u_j)] + \frac{4}{M} \sum_{k=1}^{m} \sum_{j \in I_k} [f(u_{j+1}) - f(u_j)]$$

$$\leq \frac{8}{M} \sum_{j=1}^{N-1} [f(u_{j+1}) - f(u_j)] \leq \frac{8}{M} \|f\|_{BV}.$$

If f is of bounded variation, we may, by Jordan's decomposition, represent f as the difference of two increasing functions, $f = f_1 - f_2$, such that

$$\|f\|_{BV} = \|f_1\|_{BV} + \|f_2\|_{BV}.$$

Then the hypothesis (1.40) means that

$$\left| \left[\left(\int_{-\infty}^{x_k - \delta_k} + \int_{x_k + \delta_k}^{\infty} \right) \frac{df_1(t)}{x_k - t} \right] \right.$$
$$\left. - \left[\left(\int_{-\infty}^{x_k - \delta_k} + \int_{x_k + \delta_k}^{\infty} \right) \frac{df_2(t)}{x_k - t} \right] \right| > M > 0. \qquad (1.42)$$

Thus, one of the two terms in the brackets must be greater in the absolute value than $\frac{M}{2}$, and since f_1 and f_2 are now increasing, we may apply our foregoing considerations in order to obtain the result. $\square$

Lemma 1.43. *The Hilbert transform of any step function vanishing outside of compact sets exists except at a finite number of points.*

Proof. Follows immediately from Example 1.23. $\square$

We are now in a position to prove Theorem 1.33.

Proof. Let f be real-valued. We first prove the existence a.e. of $\mathcal{H}df(x)$. By Cauchy's criterion, it is sufficient to show that, given $\varepsilon > 0$, for every x except in a set of measure less than ε

$$\left| \left(\int_{x-\delta}^{x-\delta'} + \int_{x+\delta'}^{x+\delta} \right) \frac{df(t)}{x-t} \right| \leq \varepsilon \qquad (1.44)$$

for all sufficiently small δ and δ', $0 < \delta' < \delta$.

Let $f(x) = f_{AC}(x) + f_{js}(x)$ be the Lebesgue decomposition of f into its absolutely continuous part f_{AC} and its jumps+singular part f_{js}, with

$$f_{js} = f_{jump} + f_{sing},$$

see (1.6), and let

$$f_{AC}(x) = \int_{-\infty}^{x} g(t)\,dt$$

with $g \in L^1$. Since the step functions which vanish outside of compact sets are dense in L^1, we may, to given $\varepsilon' > 0$, choose a function h such that $\|g - h\|_1 < \varepsilon'$. Thus, if

$$f_1(x) = \int_{-\infty}^{x} h(t)\,dt,$$

then $\|f_{AC} - f_1\|_{BV} < \varepsilon'$. Furthermore, f_{js} can be approximated to within ε' by a singular function f_2 with variation confined to a closed set of measure zero, that

1.3. Hilbert transform

is, which is constant on the intervals of an open set B whose complement has measure zero: $\|f_{js} - f_2\|_{BV} < \varepsilon'$. Thus, taking $\varepsilon' = \frac{\varepsilon^2}{384}$, we set $f = f_1 + f_2 + f_3$ with $\|f_3\|_{BV} < \frac{\varepsilon^2}{192}$.

Let E_ε be the set of x for which the inequality

$$\left|\left(\int_{x-\delta}^{x-\delta'} + \int_{x+\delta'}^{x+\delta}\right)\frac{df_3(t)}{x-t}\right| \leq \frac{\varepsilon}{3} \tag{1.45}$$

fails to hold for all sufficiently small δ and δ', $0 < \delta' < \delta$. Then, for every $x \in E_\varepsilon$,

$$\left|\left(\int_{-\infty}^{x-\eta} + \int_{x+\eta}^{+\infty}\right)\frac{df_3(t)}{x-t}\right| > \frac{\varepsilon}{6} \tag{1.46}$$

for sufficiently small η. By Vitali's theorem (see, e.g., [151, Ch.III, §8]) a sequence of disjoint intervals $(x_k - \eta_k, x_k + \eta_k)$ satisfying (1.46) can be chosen so as to cover E_ε, except for a set of measure zero. Then by Lemma 1.39

$$\text{meas}\{E_\varepsilon\} \leq 2\sum_k \eta_k \leq 2\frac{16}{\frac{\varepsilon}{6}}\|f_3\|_{BV} < \varepsilon.$$

Since f_1 is the integral of a step function h which vanishes outside of a compact set, by Lemma 1.43, its Hilbert-Stieltjes transform, i.e., the Hilbert transform of h, exists except for a finite number of points which we add to E_ε. Since f_2 is constant on the intervals of B, its Hilbert-Stieltjes transform obviously exists except on the complement of B which we add to E_ε. Thus, if x is not in the enlarged E_ε, there is a $\delta_0 > 0$ such that for all δ and δ', $0 < \delta' < \delta < \delta_0$, (1.45) holds for f_1, f_2, and f_3, and hence (1.44) holds as was to be shown.

The relation (1.34) now follows immediately from Lemma 1.39. In fact, if x is such that $\mathcal{H}df(x)$ exists and $|\mathcal{H}df(x)| > M$, then

$$\left|\left(\int_{-\infty}^{x-\delta} + \int_{x+\delta}^{+\infty}\right)\frac{df(t)}{x-t}\right| > \pi M \tag{1.47}$$

for sufficiently small $\delta > 0$. Again Vitali's theorem gives a sequence of disjoint intervals $(x_k - \delta_k, x_k + \delta_k)$ satisfying (1.47) which covers the set of x with $|\mathcal{H}df(x)| > M$ up to a set of measure zero. Therefore

$$\text{meas}\{x : |\mathcal{H}df(x)| > M\} \leq 2\sum_k \eta_k \leq 2\frac{16}{M}\|f\|_{BV}.$$

If f is complex-valued, then the above considerations apply to the real and imaginary part of f, respectively. Thus Theorem 1.33 is completely established. □

Theorem 1.48. *The Hilbert transform $\mathcal{H}f(x)$ of a function $g \in L^p$, $1 \leq p < \infty$, exists for almost every x.*

Proof. The case $p = 1$ follows by Theorem 1.33 since the Hilbert transform may be regarded as the Hilbert-Stieltjes transform $\mathcal{H}df$ of the absolutely continuous function
$$f(x) = \int_{-\infty}^{x} g(t)\, dt.$$
If $1 < p < \infty$, suppose $x \in [-a, a]$, $a > 0$, and set $f = f_1 + f_2$ with $f_1(t) = \chi_{[-2a, 2a]}(t) f(t)$. Then
$$\mathcal{H}_\delta f(x) = \frac{1}{\pi} \int_{|x-t| \geq \delta} \frac{f_1(t)}{x - t}\, dt$$
$$+ \frac{1}{\pi} \int_{|x-t| \geq \delta} \frac{f_2(t)}{x - t}\, dt = I_1 + I_2.$$
But
$$\lim_{\delta \to 0+} I_2 = \frac{1}{\pi} \int_{|t| \geq 2a} \frac{f(t)}{x - t}\, dt$$
for all $x \in [-a, a]$, and since $f_1 \in L^1$, $\lim_{\delta \to 0+} I_1$ exists almost everywhere. Thus the Hilbert transform of g exists for almost all $x \in [-a, a]$, and since a is arbitrary, the theorem is established. $\square$

1.3.3 Integrability of the Hilbert transform

In most of situations, we shall consider the Hilbert transform $\mathcal{H}g$ of an integrable function g. We have already proved that it exists almost everywhere. Further, if g is integrable, its Hilbert transform $\mathcal{H}g$ is not necessarily integrable. However, when it is, g has mean zero (or possesses the cancelation property)

$$\int_{\mathbb{R}} g(t)\, dt = 0. \tag{1.49}$$

This was apparently first mentioned in [101].

Let us consider a few examples, with an eye on whether the Hilbert transform is integrable or not. The first one can be found in almost every source on the subject (see, e.g., [100, p.9]), while the rest are less known.

Example 1.50. The Hilbert transform of $g(t) = \frac{1}{1+t^2}$ on the real line is given by $\mathcal{H}g(x) = \frac{x}{1+x^2}$.

We have
$$\mathcal{H}g(x) = \lim_{\delta \to 0+} \frac{1}{\pi} \int_{|x-t| \geq \delta} \frac{1}{(x-t)(1+t^2)}\, dt.$$
It is easy to check that
$$(1+x^2) \frac{1}{(x-t)(1+t^2)} = \frac{1}{x-t} + \frac{t}{1+t^2} + \frac{x}{1+t^2}.$$

1.3. Hilbert transform

The first summand on the right vanishes because of Example 1.22. Integration of the second one in the principal value sense gives zero because of the oddness of the integrand. Integrating of the third one leads to

$$\mathcal{H}g(x) = \frac{x}{\pi(1+x^2)} \lim_{\delta \to 0+} \int_{|x-t| \geq \delta} (\arctan t \,|_{-\infty}^{x-\delta} + \arctan t \,|_{x+\delta}^{\infty}) \, dt$$
$$= \frac{x}{1+x^2}. \tag{1.51}$$

The next example of the Hilbert transform, also non-integrable, can be found in [186, 5.14].

Example 1.52. Let

$$g(t) = \begin{cases} \frac{1}{t \ln^2 t}, & t > 0, \\ 0, & t \leq 0. \end{cases}$$

Then for $x > 0$,

$$|\mathcal{H}g(-x)| = \frac{1}{\pi} \int_{\mathbb{R}} \frac{g(t)}{t+x} \, dt > \frac{1}{\pi} \int_0^x \frac{dt}{2xt \ln^2 t} = \frac{1}{2\pi x \ln x}.$$

Obviously, $\mathcal{H}g \notin L^1(\mathbb{R})$.

Below we will give a similar example with a function also satisfying (1.49) but bounded.

Example 1.53. There exists an odd function with non-integrable Hilbert transform: take

$$g(t) = (t-1)^{-1} |\ln^{-2}(t-1)|$$

on $(1, \frac{3}{2})$, $g(t) = -g(-t)$ on $(-\frac{3}{2}, -1)$, and 0 otherwise. Then for $x \in (\frac{1}{2}, 1)$

$$|\mathcal{H}g(x)| \geq \left| \int_1^{1+(1-x)} \frac{1}{(t-1) \ln^2(t-1)} \frac{dt}{t-x} \right| - \frac{2}{3 \ln 2}$$
$$\geq \frac{1}{2(1-x)|\ln(1-x)|} - \frac{2}{3 \ln 2},$$

which is obviously non-integrable.

This example destroys the idea that an odd integrable function has the Hilbert transform automatically integrable, since always satisfies (1.49). Surprisingly, this example explicitly appeared in the literature as recently as in 2010, see [136].

Similarly, an example in the even case is a modification of Pitt's example given in [101, Thm.1 (b)].

Example 1.54. Take
$$g_1(t) = \frac{1}{t \ln^2 t}$$
on $[2, \infty)$ and zero otherwise, and $g_2(t) = \frac{2}{\ln 2}$ in $(0, \frac{1}{2})$ and zero otherwise. Let $g(t) = g_1(t) - g_2(t)$.

This function satisfies (1.49), is integrable on $\mathbb{R}$ and, by routine calculations as above, its Hilbert transform does not belong to $L^1(-\frac{1}{2}, 0)$. It remains to extend evenly and take into account that the even extension possesses the same properties (see, t.g., [64, Ch.III, Lemma 7.40, p. 354]).

Though we are not going to use it, let us mention that the Hilbert transform of an integrable function g always satisfies the so-called weak or Kolmogorov type estimate, to wit (cf. (1.34))

$$\operatorname{meas}\{x : |\mathcal{H}g(x)| > M\} \lesssim \frac{1}{M} \|f\|_{L^1(\mathbb{R})}. \tag{1.55}$$

1.3.4 Special cases of the Hilbert transform

When in the definition of the Hilbert transform (1.66) the function g is odd, we will denote this transform by $\mathcal{H}_o$, and by simple substitution it is equal to

$$\mathcal{H}_o g(x) = \frac{2}{\pi} \int_0^\infty \frac{t g(t)}{x^2 - t^2} \, dt. \tag{1.56}$$

Correspondingly, when in the definition of the Hilbert transform (1.66) the function g is even, we will denote this transform by $\mathcal{H}_e$, and it is equal to

$$\mathcal{H}_e g(x) = \frac{2}{\pi} \int_0^\infty \frac{x g(t)}{x^2 - t^2} \, dt. \tag{1.57}$$

There exist clear relations between the even and odd Hilbert transforms. Here and in what follows we use the notations "$\lesssim$" and "$\gtrsim$" as abbreviations for "$\leq C$" and "$\geq C$", with C being an absolute positive constant.

Proposition 1.58. *Let $g \in L^1(\mathbb{R}_+)$. Then*

$$\mathcal{H}_e g(x) = \mathcal{H}_o g(x) + \frac{2}{\pi x} \int_0^x g(t) \, dt + G(x), \tag{1.59}$$

where

$$\int_0^\infty |G(x)| \, dx \lesssim \int_0^\infty |g(t)| \, dt.$$

Proof. The first step is as follows:

$$\mathcal{H}_e g(x) = \mathcal{H}_o g(x) + \frac{2}{\pi} \int_0^\infty \frac{g(t)}{x+t} \, dt. \tag{1.60}$$

1.3. Hilbert transform

Rewriting the last integral as

$$\int_0^\infty \frac{g(t)}{x+t}\,dt = \frac{1}{x}\int_0^x g(t)\,dt$$
$$-\int_0^x g(t)\frac{t}{x(x+t)}\,dt + \int_x^\infty \frac{g(t)}{x+t}\,dt,$$

we can check that the last two summands on the right are integrable, as required. More details will be given in the proofs of the main theorems in the next chapter, where (1.59) will be used in concrete situations. □

The difference between the two Hilbert transforms in (1.60) only looks completely symmetric; their interaction with $\cos xt$ and $\sin xt$ differs quite seriously one from the other.

We will see in what follows that the well-known extension of Hardy's inequality (see, e.g., [64, (7.24)])

$$\int_\mathbb{R} \frac{|\widehat{g}(x)|}{|x|}\,dx \lesssim \|g\|_{H^1(\mathbb{R})} \quad (1.61)$$

plays a crucial role in these considerations. The space $H^1(\mathbb{R})$ will be defined later, but it is worth mentioning that here and in most of situations it will mainly mean that both $g, \mathcal{H}g \in L^1(\mathbb{R})$. To get a flavor of it, recall that for $g \in L^1(\mathbb{R})$ only, $\widehat{g}$ may have an arbitrarily slow decay near infinity. Here (1.61) shows that if, in addition, $\mathcal{H}g \in L^1(\mathbb{R})$, then the Fourier transform decays faster. Since the name "Hardy's inequality" can refer to two different well-known inequalities, we dare to introduce a new name for (1.61) and its versions. Since (1.61) also involves an important aspect of the Fourier transform, we shall call it the *Fourier–Hardy inequality*, as is already mentioned in the introduction. Recall that the other inequality, the one most people conceive of when hearing the words "Hardy's inequality", is (see, e.g., [79, (330)]):

$$\left[\int_0^\infty |x^{2b}R(x)|^2\,dx\right]^{\frac{1}{2}} \lesssim \left[\int_0^\infty x^{2b+2}\psi(x)^2\,dx\right]^{\frac{1}{2}}, \quad (1.62)$$

where, for $\psi(x) \geq 0$, either

$$R(x) = \int_0^x \psi(t)\,dt$$

and $2b < -1$, or

$$R(x) = \int_x^\infty \psi(t)\,dt$$

and $2b > -1$. There are more advanced versions, for different metrics and various weights.

Let f be a function with bounded variation on $\mathbb{R}$, vanishing at infinity. Since it need not be integrable, its Hilbert transform, a usual substitute for the conjugate function, may not exist. One has to use the modified Hilbert transform as the conjugate of a bounded function (see, e.g., [102, (3.1)] or [65, Ch.III, §1])

$$\widetilde{\mathcal{H}}f(x) = (\text{P.V.})\frac{1}{\pi}\int_{\mathbb{R}} f(t)\left\{\frac{1}{x-t} + \frac{t}{1+t^2}\right\} dt. \quad (1.63)$$

This modification was introduced by Kober in [102] for L^∞ functions. Such a function is locally in any L^p; therefore it is expected to behave in regard to singularity like any L^p function. And indeed this is the case for $\widetilde{\mathcal{H}}$, since the problems near infinity are balanced by that term $\frac{t}{1+t^2}$. More precisely, the difference

$$\widetilde{\mathcal{H}}f(x) - \mathcal{H}f(x) = \int_{\mathbb{R}} f(t)\frac{t}{1+t^2} dt,$$

which is bounded for any L^p function f, $1 \leq p < \infty$, by Hölder's inequality, since the function $\frac{t}{1+t^2}$ belongs to $L^{p'}$. If $f \in L^\infty$, one just has to consider separately a neighborhood of the singularity, where any L^p argument for $1 \leq p < \infty$ works. In the sense of complex analysis, this is a conjugate function exactly as without this term. In fact, in most situations one can forget about the initial definition of the Hilbert transform and always use (1.63). Indeed, the kernel

$$\frac{1}{x-t} + \frac{t}{1+t^2}$$

appeared, in a sense, at least 20 years earlier. In many sources it is attributed to R. Nevanlinna [153] (see, e.g., [106, Appendix] or [5, Ch. III, §1]). Without details, the well-known Riesz–Herglotz kernel

$$\frac{e^{i\theta}+z}{e^{i\theta}-z},$$

where $\theta \in [-\pi, \pi]$ and z lies in the unit disk, reduces to it, up to a constant and with z in place of x, by substituting $-\cot\frac{\theta}{2} = t$ and replacing z with $\frac{z-i}{z+i}$. The latter just maps the unit disk in $\mathbb{C}$ to the upper half-plane. Correspondingly, integration over the unit circle turns to integration over $\mathbb{R}$. Also, the Riesz–Herglotz kernel becomes $\frac{t-i}{t+i}$. More precisely, in [153] the kernel never appears in the form we use. However, how to pass to it from the equivalent form it was given there is well known. This was first done in [39].

Concerning (1.29), we would like to mention the following relations for the cosine and sine Fourier transforms. We have

$$\widehat{f_s}(x) = \mathcal{H}\widehat{f_c}(x). \quad (1.64)$$

and

$$\widehat{f_c}(x) = -\mathcal{H}\widehat{f_s}(x). \quad (1.65)$$

The formulas (1.64) and (1.65) are known, see [100, (5.42) and (5.43)] for square integrable functions. For more delicate use of these formulas in the L^1 setting, see [117].

1.3. Hilbert transform

Similar notions exist in the discrete setting. For the sequence $a = \{a_k\} \in \ell^1$, the discrete Hilbert transform is defined for $m \in \mathbb{Z}$ as (see, e.g., [100, (13.127)], [30])

$$\hbar a(m) = \sum_{\substack{k=-\infty \\ k \neq m}}^{\infty} \frac{a_k}{m-k}. \tag{1.66}$$

If the sequence a is either even or odd, the corresponding Hilbert transforms $\hbar_e$ and $\hbar_o$ may be expressed in a special form (see, e.g., [8] or [100, (13.130) and (13.131)]). More precisely, if a is even, with $a_0 = 0$, we have $\hbar_e(0) = 0$ and for $m = 1, 2, \ldots$

$$\hbar_e a(m) = \sum_{\substack{k=1 \\ k \neq m}}^{\infty} \frac{2m a_k}{m^2 - k^2} + \frac{a_m}{2m}. \tag{1.67}$$

If a is odd, with $a_0 = 0$, we have for $m = 0, 1, 2, \ldots$

$$\hbar_o a(m) = \sum_{\substack{k=1 \\ k \neq m}}^{\infty} \frac{2k a_k}{m^2 - k^2} - \frac{a_m}{2m}. \tag{1.68}$$

Of course, $\frac{a_0}{0}$ is considered to be zero.

The discrete Hardy spaces will be denoted by h, h_e and h_o and will mean, for an ℓ^1 sequence a, that the corresponding discrete Hilbert transform is also in ℓ^1.

1.3.5 Conditions for the integrability of the Hilbert transform

The above results naturally give rise to the problem of conditions which ensure the integrability of the Hilbert transform. To begin with, the following simple statement (see [175, Ch.2, §7]) holds.

Proposition 1.69. *Suppose g is a bounded function on $\mathbb{R}$ with compact support. Then $\mathcal{H}g \in L^1(\mathbb{R})$ if and only if g satisfies (1.49).*

Proof. We have

$$\mathcal{H}g(x) = \frac{\widehat{g}(0)}{\pi x} + O\left(\frac{1}{x^2}\right)$$

as $|x| \to \infty$. □

For example, this simple test allows one to characterize the functions from the Schwartz class whose Hilbert transform is integrable. This happens if and only if (1.49) holds true.

On the other hand, one of the most general conditions of this kind, the so-called Zygmund–Stein condition, should be mentioned (see, e.g., [171]). It is connected to the so-called class $L \log L$, which contains all the functions satisfying

$$\int_{\mathbb{R}} |f(t)| \ln^+ |f(t)| \, dt < \infty,$$

where $\ln^+ A$ means $\ln A$ if $A > 1$ and zero otherwise. Roughly speaking, that condition in its "Zygmund part", the sufficient condition, says that a function with compact support, being in $L \log L$ and satisfying (1.49), has integrable Hilbert transform. Its necessary, "Stein side" says, with certain abuse of formality, that a function with integrable Hilbert transform is in $L \log L$ on the interval where it is like-sign.

Let us give a different condition of that sort, that is practical.

Theorem 1.70. *Suppose g is a bounded function on $\mathbb{R}$ such that for some non-negative function φ*

$$|g(t)| \leq \frac{\varphi(t)}{1+|t|} \tag{1.71}$$

with

$$\sum_{k=0}^{\infty} k\lambda_k = \sum_{k=0}^{\infty} k \sup_{2^{k-1} < |t| \leq 2^k} \varphi(t) < \infty. \tag{1.72}$$

If it has mean zero, then $\mathcal{H}g \in L^1(\mathbb{R})$.

We postpone the proof of this test, since it will be proved within the scope of the real Hardy space.

This is a generalization of the result in [175, Ch.2, §7] where $g(t) = O(\frac{1}{1+t^2})$. Roughly speaking, a bounded function with mean zero and appropriate smoothness near infinity belongs to the Hardy space, not depending on whether it is general, odd or even. A representative example of a condition that satisfies Theorem 1.70 is a function that behaves as $g(t) = O\left(\frac{1}{t \ln^\alpha t}\right)$, with $\alpha > 2$.

To give an idea of the line that one cannot cross, or, in other words, to show that $\alpha > 2$ is a sharp condition in the above example, we will present an example of an even function g, which satisfies (1.49) and whose Hilbert transform is not integrable. For a similar Pitt's example; see Example 1.54 or [101, Thm.1(b)].

Example 1.73. Take $g_1(t) = \frac{1}{t \ln^2 t}$ on $[e, \infty)$ and zero otherwise, and $g_2(t) = 1$ in $(0, 1)$ and zero otherwise. Let $g(t) = g_1(t) - g_2(t)$. This function satisfies (1.49), and is integrable on $\mathbb{R}$. Consider its Hilbert transform times π for $x \in (-\infty, -2)$:

$$\int_0^1 \frac{dt}{|x|+t} + \int_e^\infty \frac{dt}{t(|x|+t)\ln^2 t}$$

$$= \frac{1}{|x|} \int_e^\infty \frac{dt}{(|x|+t)\ln^2 t} - \frac{1}{|x|} + \ln \frac{|x|+1}{|x|}$$

$$\geq \frac{1}{(|x|+e)\ln(|x|+e)} - \frac{1}{|x|} + \ln \frac{|x|+1}{|x|}.$$

Since the last two terms considered together behave as $O(\frac{1}{|x|^2})$ for large $|x|$, the Hilbert transform behaves as

$$\frac{1}{(|x|+e)\ln(|x|+e)}$$

1.4. Hardy spaces and subspaces

for the considered x and thus is not integrable. What remains is to extend it to become even and take into account that the even extension possesses the same integrability properties (see [64, Ch.III, Lemma 7.40, p. 354]).

Remark 1.74. One direction in the latter property is almost immediate. More precisely, let g be an integrable function on $\mathbb{R}_+$. If the Hilbert transform of it extended as zero to $(-\infty, 0)$, written g_0, is integrable over $\mathbb{R}$, then the Hilbert transform of its even extension g_e is integrable over $\mathbb{R}$. Indeed, by (1.67),

$$\mathcal{H}g_e(x) = \mathcal{H}_e g(x) = \frac{2}{\pi} \int_0^\infty \frac{xg(t)}{x^2 - t^2} dt$$
$$= \frac{1}{\pi} \int_0^\infty \frac{g(t)}{x - t} dt + \frac{1}{\pi} \int_0^\infty \frac{g(t)}{x + t} dt. \quad (1.75)$$

If $\mathcal{H}g_0 \in L^1(\mathbb{R})$, it remains to observe that

$$\int_{-\infty}^0 \left| \int_0^\infty \frac{g(t)}{x - t} dt \right| dx = \int_0^\infty \left| \int_0^\infty \frac{g(t)}{x + t} dt \right| dx$$

and

$$\int_0^\infty |\mathcal{H}g_e(x)| dx \leq \frac{1}{\pi} \int_0^\infty \left| \int_0^\infty \frac{g(t)}{x - t} dt \right| dx$$
$$+ \frac{1}{\pi} \int_0^\infty \left| \int_0^\infty \frac{g(t)}{x + t} dt \right| dx = \|\mathcal{H}g_0\|_{L^1(\mathbb{R})}.$$

The other direction is proved in [64, Ch.III, Lemma 7.40, p. 354] by means of atomic characterization.

It will be shown in Subsection 1.4.4 that the Hilbert transform of an odd function may behave in a completely different way.

1.4 Hardy spaces and subspaces

As mentioned, if g is integrable, its Hilbert transform is not necessarily integrable. When it is, we say that g is in *the (real) Hardy space* $H^1 := H^1(\mathbb{R})$. Among a variety of its characterizations (or, equivalently, definitions), we shall mainly deal with the one given by means of the Hilbert transform.

Definition 1.76. The real Hardy space $H^1 := H^1(\mathbb{R})$ is the subspace of $L^1(\mathbb{R})$ which consists of functions with integrable Hilbert transform. It is endowed with the norm

$$\|g\|_{H^1(\mathbb{R})} = \|g\|_{L^1(\mathbb{R})} + \|\mathcal{H}g\|_{L^1(\mathbb{R})}. \quad (1.77)$$

If $g \in H^1(\mathbb{R})$, then, as mentioned above, (1.49) holds true. The functions $g \in L^1(\mathbb{R})$, which satisfy (1.49), are sometimes called wavelet functions (see, e.g.,

[164]), and the corresponding space is denoted by $L_0^1(\mathbb{R})$. By this, strictly speaking, $H^1(\mathbb{R})$ is a subspace of $L_0^1(\mathbb{R})$, which, in turn, is a subspace of $L^1(\mathbb{R})$. For example, $H^1(\mathbb{R})$ is dense in $L_0^1(\mathbb{R})$ but not in the whole $L^1(\mathbb{R})$.

An odd function always satisfies (1.49). However, not every odd integrable function belongs to $H^1(\mathbb{R})$; for counterexamples, see [136] (or Example 1.53) and [114].

For functions in $L^p(\mathbb{R})$, $1 < p < \infty$, the situation is completely different. A deep theorem due to M. Riesz asserts that H^p spaces for these p, in fact, coincide with L^p. More precisely

$$\|\mathcal{H}g\|_{L^p(\mathbb{R})} \leq C_p \|g\|_{L^p(\mathbb{R})}. \tag{1.78}$$

Here C_p is the well-known Pichorides (Gohberg–Krupnik for $p = 2^k$, see, e.g., [98]) constant:

$$C_p = \begin{cases} \tan \frac{\pi}{2p}, & 1 < p \leq 2, \\ \cot \frac{\pi}{2p}, & 2 \leq p < \infty. \end{cases}$$

It is clear that the constant C_p depends on p in such a way that $\lim_{p \to 1} C_p = \infty$. The same, of course, occurs if $p \to \infty$.

Recall that we are going to mostly concentrate on functions on the half-axis. This also is about the Hardy space if we consider only odd or only even functions.

Definition 1.79. If the odd Hilbert transform is integrable, we shall denote the corresponding Hardy space by $H_o^1(\mathbb{R}_+)$, or sometimes simply H_o^1. Symmetrically, if the even Hilbert transform is integrable, we shall denote the corresponding Hardy space by $H_e^1(\mathbb{R}_+)$, or sometimes simply H_e^1.

Clearly, to have (1.49) for the latter class reduces to

$$\int_0^\infty g(t)\, dt = 0. \tag{1.80}$$

1.4.1 Atomic characterization

The real Hardy space can be characterized in a different way than (1.77), via the so-called atomic decomposition. Let $a(x)$ denote an atom (a $(1, p, 0)$-atom), $1 < p \leq \infty$, a function that is of compact support:

$$\operatorname{supp} a \subset I = [c, d], \quad -\infty < c < d < +\infty; \tag{1.81}$$

and satisfies the following size condition (L^p normalization), with $|I| = d - c$,

$$\|a\|_p \leq \frac{1}{|I|^{1-\frac{1}{p}}}; \tag{1.82}$$

1.4. Hardy spaces and subspaces

and the cancelation condition

$$\int_{\mathbb{R}} a(x)\, dx = 0. \tag{1.83}$$

It is well known (see, e.g., [173, Ch.III, 2.2]; also [64] or [175]) that

$$\|f\|_{H^1} \sim \inf\left\{\sum_k |\lambda_k| < \infty : f(x) = \sum_k \lambda_k a_k(x)\right\}, \tag{1.84}$$

where a_k are the above described atoms, with certain p fixed, and $\sum_k |\lambda_k| < \infty$ ensures that the sum $\sum_k \lambda_k a_k(x)$ converges in the L^1 norm. Any fixed $p > 1$ leads to the same space, which is equivalent to the real Hardy space defined above via the Hilbert transform. Taking for each k its p_k leads to new, wider spaces provided $p_k \to 1$ (see [177]) or, under some additional assumptions, to the atomic characterization of $\mathbb{R}$ (see [1] and [2]).

An obvious advantage of this characterization is that in many situations it suffices to verify a certain statement only on atoms. However, this is not a "law of Nature" and one should be cautious in using this approach; see a celebrated example due to Bownik in [29].

What is of crucial importance for us is that all the characterizations of the real Hardy space are equivalent. This means, in particular, that proving a statement by means of atomic decomposition we have the same fact in terms of (1.77).

Despite of the above warning, in most situations atomic characterization works pretty well. To illustrate this, let us make use of it for proving (1.61). It will be proved if we show that for any atom a there holds

$$\int_{\mathbb{R}} \frac{|\widehat{a}(x)|}{|x|}\, dx \leq C, \tag{1.85}$$

with C independent of the atom. Let a be a $(1, p, 0)$-atom, $1 < p \leq 2$, and let us split the integral on the left-hand side into two: over $|x| \leq \frac{1}{|I|}$ and over $|x| > \frac{1}{|I|}$. For the first one, we have, by (1.83), Hölder's inequality and (1.82),

$$\int_{|x| \leq \frac{1}{|I|}} \frac{|\widehat{a}(x)|}{|x|}\, dx = \int_{|x| \leq \frac{1}{|I|}} \frac{1}{|x|} \left|\int_c^d a(t)[e^{-ixt} - e^{-ixa}]\, dt\right| dx$$

$$\leq \int_{|x| \leq \frac{1}{|I|}} \frac{1}{|x|} \|a\|_{L^p} \left(\int_c^d |e^{-ixt} - e^{-ixa}|^{p'} dt\right)^{\frac{1}{p'}} dx$$

$$\leq \int_{|x| \leq \frac{1}{|I|}} |I|^{\frac{1}{p}-1} \left(\int_c^d (t-a)^{p'} dt\right)^{\frac{1}{p'}} dx$$

$$\leq \frac{2}{(p'+1)^{\frac{1}{p'}}}.$$

For the second integral, applying Hölder's inequality and the Hausdorff–Young inequality, we obtain

$$\int_{|x|>\frac{1}{|T|}} \frac{|\widehat{a}(x)|}{|x|} dx \leq \left(\int_{|x|>\frac{1}{|T|}} |x|^{-p} dx\right)^{\frac{1}{p}} \left(\int_{\mathbb{R}} |\widehat{a}(x)|^{p'} dx\right)^{\frac{1}{p'}}$$

$$\leq \frac{1}{(p-1)^{\frac{1}{p}}} (d-c)^{\frac{1}{p}-1} \left|\int_c^d |a(t)|^p dt\right|^{\frac{1}{p}}.$$

Using (1.82) for the last integral, we arrive at the required bound.

One more illustration comes from the postponed

Proof of Theorem 1.70. The scheme of the proof is the same as in [175, Ch.2, §7] for the case where $g(t) = O(\frac{1}{1+t^2})$. We denote $g_0(t) = g(t)$ when $|t| \leq 1$ and $g_0(t) = 0$ otherwise. Further, for $k = 1, 2, \ldots$ define

$$g_k(t) = \begin{cases} g(t), & 2^{k-1} < |t| \leq 2^k, \\ 0, & \text{otherwise} \end{cases}$$

and

$$c_k = \int_{\mathbb{R}} g_k(t)\, dt = \int_{2^{k-1}<|t|\leq 2^k} g(t)\, dt.$$

Note that under the assumptions of the theorem g is an integrable function, since

$$|c_k| \leq \int_{2^{k-1}<|t|\leq 2^k} \frac{\lambda_k}{t}\, dt. \tag{1.86}$$

We have

$$g(t) = \sum_{k=0}^{\infty} g_k(t).$$

Let us also denote

$$S_k = \sum_{j \geq k} c_j;$$

it follows from the assumption of the theorem that $S_0 = 0$.

Taking a bounded function η supported in $\{|t| \leq 1\}$ and such that

$$\int_{\mathbb{R}} \eta(t)\, dt = 1$$

and denoting $\eta_k(t) = \frac{1}{2^k}\eta(\frac{t}{2^k})$, we obtain

$$\int_{\mathbb{R}} \eta_k(t)\, dt = 1$$

1.4. Hardy spaces and subspaces

and
$$g(t) = \sum_{k=0}^{\infty}[g_k(t) - c_k\eta_k(t)] + \sum_{k=0}^{\infty} c_k\eta_k(t). \tag{1.87}$$

The first sum is
$$\sum_{k=0}^{\infty}[g_k(t) - c_k\eta_k(t)] = \sum_{k=0}^{\infty} \frac{1}{2^k} \int_{2^{k-1}<|x|\leq 2^k} [g_k(t) - g(x)\eta(\frac{t}{2^k})]\,dx$$
$$= \sum_{k=0}^{\infty} B_k(t).$$

We have
$$\int_{\mathbb{R}} B_k(t)\,dt = \int_{2^{k-1}<|x|\leq 2^k} g(x)\,dx \left[1 - \int_{\mathbb{R}} \eta_k(t)\,dt\right] = 0.$$

The function $B_k(t)$ is supported in $\{|t| \leq 2^k\}$. Finally, for $2^{k-1} < |t| \leq 2^k$, there holds $|g_k(t)| \leq C\frac{\lambda_k}{2^k}$. Combining this with (1.86), we obtain
$$\|B_k(t)\|_\infty \leq C\frac{\lambda_k}{2^k}.$$

By this, $B_k(t) = C\lambda_k A_k(t)$, where $A_k(t)$ is an atom supported in $\{|t| \leq 2^k\}$. Here, the numbers λ_k are the coefficients of an atomic decomposition of g.

Concerning the second sum in (1.87), it can be represented in an equivalent form due to the mean zero property ($S_0 = 0$):
$$\sum_{k=1}^{\infty} S_k(\eta_k - \eta_{k-1}).$$

Each value $\eta_k(t) - \eta_{k-1}(t)$ is a multiple of an analogous atom. The numbers S_k are appropriate for an atomic decomposition, since
$$\sum_{j=0}^{\infty} |S_j| = \sum_{j=0}^{\infty} \left|\sum_{k=j}^{\infty} c_k\right| \leq \sum_{j=0}^{\infty}\sum_{k=j}^{\infty} \int_{2^{k-1}<|t|\leq 2^k} |g(t)|\,dt$$
$$\leq C\sum_{j=0}^{\infty}\sum_{k=j}^{\infty} \lambda_k = C\sum_{k=1}^{\infty} k\lambda_k,$$

and (1.72) completes the proof. □

1.4.2 Molecular characterization

The idea of molecular decompositions is due to Coifman and Weiss [45]; a good description can be found in [64, Ch.III, §7]. Recall that a function M is called a molecule (centered at $x_0 \in \mathbb{R}$) if it satisfies (1.49) and

$$\left(\int_{\mathbb{R}} |M(x)|^2 dx\right)^{\frac{1}{4}} \left(\int_{\mathbb{R}} |x-x_0|^2 |M(x)|^2 dx\right)^{\frac{1}{4}} < \infty.$$

The left-hand side is called a molecular norm of M and is denoted by $N(M)$. Every $(1,2,0)$ atom $a(t)$ is a molecule. Indeed,

$$\int_c^d |a(t)|^2 (t-c)^2 dt \leq (d-c),$$

and together with (1.82), with $p = 2$, this implies the finiteness of the molecular norm.

A handy machinery is delivered by the following fact (see, e.g., [64, Ch.III, §7, p.328]).

Proposition 1.88. *A function g belongs to $H^1(\mathbb{R})$ if and only if*

$$g(x) = \sum_{j=1}^{\infty} M_j(x)$$

for almost all x, with the M_j-s being molecules which satisfy

$$\sum_{j=1}^{\infty} N(M_j) < \infty.$$

To get an idea of how molecules work, let us prove that such an M without satisfying (1.49) is Lebesgue integrable. Adding (1.49) implies that it belongs to $H^1(\mathbb{R})$. We follow [64, Ch.III, §7].

The former is very simple ([64, Ch.III, §7, Lemma 7.11]). Indeed, applying the Cauchy–Schwarz–Bunyakovskii inequality, we get

$$\int_{\mathbb{R}} |M(t)|\, dt = \left(\int_{|t| \leq r} + \int_{|t| > r}\right) |M(t)|\, dt$$

$$\leq \sqrt{2r}\|M\|_{L^2(\mathbb{R})} + \sqrt{\frac{2}{r}}\|tM(t)\|_{L^2(\mathbb{R})}.$$

Both norms are finite by definition of molecule. One can take $r = 1$ but also one can play with r in certain situations. The latter is more important, since taking

$$r = \frac{\|tM(t)\|_{L^2(\mathbb{R})}}{\|M\|_{L^2(\mathbb{R})}},$$

we obtain

$$\|M\|_{L^1(\mathbb{R})} \leq 2^{\frac{3}{2}} \|M\|_{L^2(\mathbb{R})}^{\frac{1}{2}} \|tM(t)\|_{L^2(\mathbb{R})}^{\frac{1}{2}},$$

1.4. Hardy spaces and subspaces

which gives rise to molecular characterization of $H^1(\mathbb{R})$.

Further, let (1.49) hold in addition. Since we already know that $M \in L^1(\mathbb{R})$, it suffices to prove that $\mathcal{H}M \in L^1(\mathbb{R})$ ([64, Ch.III, §7, Lemma 7.12]). We have $M \in L^2(\mathbb{R})$, hence $\mathcal{H}M \in L^2(\mathbb{R})$, more precisely,

$$\|\mathcal{H}M\|_{L^2(\mathbb{R})} \leq \|M\|_{L^2(\mathbb{R})},$$

since $C_2 = 1$. Now,

$$\begin{aligned} x\mathcal{H}M(x) &= \frac{1}{\pi} \int_{\mathbb{R}} \frac{x}{x-t} M(t)\,dt \\ &= \frac{1}{\pi} \int_{\mathbb{R}} M(t)\,dt + \frac{1}{\pi} \int_{\mathbb{R}} \frac{1}{x-t} tM(t)\,dt \\ &= \mathcal{H}(tM(t))(x). \end{aligned}$$

This yields

$$\|x\mathcal{H}M(x)\|_{L^2(\mathbb{R})} \leq \|tM(t)\|_{L^2(\mathbb{R})};$$

hence, as above,

$$\|\mathcal{H}M\|_{L^1(\mathbb{R})} \leq 2^{\frac{3}{2}} \|M\|_{L^2(\mathbb{R})}^{\frac{1}{2}} \|tM(t)\|_{L^2(\mathbb{R})}^{\frac{1}{2}}.$$

Like for atoms, there are another types of q-molecules which depend on a parameter q, with $q = 2$ in the considered case; however, they are of more importance for the H^p spaces with $p < 1$ rather than for H^1. Nevertheless, let us present the explicit value for the q-molecular norm (see, e.g., [64, Ch.III, §7]):

$$N_q(M) = \|M\|_{L^q(\mathbb{R})}^{\frac{1}{q}} \|tM(t)\|_{L^q(\mathbb{R})}^{\frac{1}{q'}}.$$

All the above reasoning goes along the same lines. Instead, let us show how (1.61) can be proved by the molecular approach. We assume $1 < q \leq 2$, and in order to estimate for the q-molecule M the integral

$$\int_{\mathbb{R}} \left| \frac{\widehat{M}(x)}{x} \right| dx,$$

we split the integral in two: over $|x| \leq r$ and over $|x| > r$, with

$$r = \frac{\|M\|_{L^q(\mathbb{R})}}{\|tM(t)\|_{L^q(\mathbb{R})}}.$$

First, applying Hölder's inequality and then the Hausdorff-Young inequality, we obtain

$$\int_{|x|>r} \left| \frac{\widehat{M}(x)}{x} \right| dx \leq \left(\int_{|x|>r} |\widehat{M}(x)|^{q'} dx \right)^{\frac{1}{q'}} \left(\int_{|x|>r} |x|^{-q} dx \right)^{\frac{1}{q}}$$

$$\leq \|M\|_{L^q(\mathbb{R})} \left(\frac{2}{q-1} \right)^{\frac{1}{q}} r^{-\frac{1}{q'}} = \left(\frac{2}{q-1} \right)^{\frac{1}{q}} N_q(M).$$

For the integral over $|x| \leq r$, we additionally split the integral in t into two: over $|t| \leq \frac{1}{|x|}$ and over $|t| > \frac{1}{|x|}$. In addition, for both we estimate the Fourier integrals with $e^{-ixt} - 1$ rather than with e^{-ixt}, by the cancelation property. Using only Hölder's inequality, we get

$$\int_{|x| \leq r} \left| \int_{|t| > \frac{1}{|x|}} tM(t) \frac{e^{-ixt} - 1}{xt} \, dt \right| dx$$

$$\leq \int_{|x| \leq r} \int_{|t| > \frac{1}{|x|}} |tM(t)| \frac{dt}{|t|} \frac{dx}{|x|}$$

$$\leq \int_{|x| \leq r} \|tM(t)\|_{L^q(\mathbb{R})} \left(\int_{|t| > \frac{1}{|x|}} |t|^{-q'} dt \right)^{\frac{1}{q'}} \frac{dx}{|x|}$$

$$= 2q \left(\frac{2}{q'-1} \right)^{\frac{1}{q'}} N_q(M).$$

Finally, by the trivial estimate $\left| \frac{e^{-ixt}-1}{xt} \right| \leq 1$ and again by Hölder's inequality,

$$\int_{|x| \leq r} \left| \int_{|t| \leq \frac{1}{|x|}} tM(t) \frac{e^{-ixt} - 1}{xt} \, dt \right| dx \leq \int_{|x| \leq r} \|tM(t)\|_{L^q(\mathbb{R})} \left(\int_{|t| \leq \frac{1}{|x|}} dt \right)^{\frac{1}{q'}} dx$$

$$= 2q 2^{\frac{1}{q'}} N_q(M).$$

Combining all these estimates, we obtain

$$\int_{\mathbb{R}} \left| \frac{\widehat{M}(x)}{x} \right| dx \leq \left[\left(\frac{2}{q-1} \right)^{\frac{1}{q}} + 2q \left(\frac{2}{q'-1} \right)^{\frac{1}{q'}} + 2q 2^{\frac{1}{q'}} \right] \|M\|_{L^q(\mathbb{R})}^{\frac{1}{q}} \|tM(t)\|_{L^q(\mathbb{R})}^{\frac{1}{q'}}.$$

With the Babenko–Beckner constant, the factor before $N_q(M)$ could be different but this does not seem essential and still is not clear whether this factor is sharp.

Again, with certain precautions one can prove many results for the whole space by checking them only on molecules. To compare the two characterizations, recall that every atom is also a molecule. On the other hand, each molecule can be represented by means of a collection of atoms.

1.4.3 Integrability spaces

When we prove Theorem 2.8, we will see that the Fourier transform of a function of bounded variation is tied up with the Hilbert transform of a related function and, correspondingly, with $H_o^1(\mathbb{R}_+)$. There is a scale of subspaces of $H_o^1(\mathbb{R}_+)$ proved to be convenient in applications.

1.4. Hardy spaces and subspaces

Definition 1.89. For $1 < q < \infty$, the space O_q is the space of functions g with finite norm

$$\|g\|_{O_q} = \int_0^\infty \left(\frac{1}{x}\int_x^{2x} |g(t)|^q dt\right)^{\frac{1}{q}} dx. \tag{1.90}$$

All these spaces and their sequence analogs first appeared in the paper by D. Borwein [28], but became – for sequences – widely known after the paper by G.A. Fomin [59]; see also [66, 67]. On the other hand, these spaces are a partial case of the so-called Herz spaces (see first of all the initial paper by Herz [86] and a relevant paper of Flett [58]; see also [56] and [199]).

Further, for $q = \infty$ we define the corresponding space as follows.

Definition 1.91. The space O_∞ is the space of functions g with finite norm

$$\|g\|_{O_\infty} = \int_0^\infty \operatorname*{ess\,sup}_{x \le t \le 2x} |g(t)|\, dx. \tag{1.92}$$

The role of the integrable monotone majorant for problems of almost everywhere convergence of singular integrals is known from the work of D.K. Faddeev (see, e.g., [6, Ch.IV, §4]; also [176, Ch.I]); for spectral synthesis problems it was used by Beurling [23], for more details see [20].

The basic interrelations between these spaces are given by the chain of embeddings

$$O_\infty \hookrightarrow O_{p_1} \hookrightarrow O_{p_2} \hookrightarrow H_o^1 \hookrightarrow L^1 \ (p_1 > p_2 > 1), \tag{1.93}$$

which can be found in [107], [92, Ch.3], [121].

To give an example of the difference (proper embedding) between these spaces (see [113]), we need certain preliminaries.

In the problems of integrability of the Fourier transform, the following T-transform of a function $g(t)$ defined on $\mathbb{R}_+ = [0, \infty)$ is of importance:

$$\int_0^{\frac{x}{2}} \frac{g(x-t) - g(x+t)}{t}\, dt = \int_{\frac{x}{2}}^{\frac{3x}{2}} \frac{g(t)}{x-t}\, dt, \tag{1.94}$$

understood in the same sense as of principal value. It is called the Telyakovskii transform in [61].

First, let us show how the T-transform and the Hilbert transform of an odd integrable function are related. This is useful in applications.

Lemma 1.95. Let g be an odd integrable function. Then for any $0 < a < 1 < b$ and $x > 0$,

$$\mathcal{H}g(x) = \frac{1}{\pi}\int_\mathbb{R} \frac{g(t)}{t-x}\, dt = \frac{1}{\pi}\int_{ax}^{bx} \frac{g(t)}{t-x}\, dt + G(x), \tag{1.96}$$

where G is such that

$$\int_0^\infty |G(x)|\, dx \le \frac{2}{\pi \min(b-1, 1-a)}\int_0^\infty |g(t)|\, dt. \tag{1.97}$$

Proof. By the oddness, we have for $x > 0$

$$\int_0^\infty \left| \left(\int_{-\infty}^{-bx} + \int_{bx}^\infty \right) \frac{g(t)}{t-x} \, dt \right| dx = 2 \int_0^\infty \left| \int_{bx}^\infty \frac{tg(t)}{t^2 - x^2} \, dt \right| dx$$

$$\leq 2 \int_0^\infty t|g(t)| \int_0^{t/b} \frac{dx}{t^2 - x^2} \, dt$$

$$\leq \frac{2b^2}{b^2 - 1} \int_0^\infty t|g(t)| \int_0^{t/b} \frac{dx}{t^2} \, dt$$

$$\leq \frac{2}{b-1} \int_0^\infty |g(t)| \, dt. \tag{1.98}$$

Similarly,

$$\int_0^\infty \left| \left(\int_{-ax}^0 + \int_0^{ax} \right) \frac{g(t)}{t-x} \, dt \right| dx \leq \frac{2}{1-a} \int_0^\infty |g(t)| \, dt. \tag{1.99}$$

Finally,

$$\int_0^\infty \left| \int_{-bx}^{-ax} \frac{g(t)}{t-x} \, dt \right| dx \leq \int_0^\infty \frac{1}{x} \int_{ax}^{bx} |g(t)| \, dt \, dx$$

$$\leq \frac{1}{\ln(b/a)} \int_0^\infty |g(t)| \, dt$$

$$\leq \frac{2}{\min(b-1, 1-a)} \int_0^\infty |g(t)| \, dt. \tag{1.100}$$

The lemma is proved. $\square$

It is clear now how the T-transform and the Hilbert transform interplay. With this in hand, let us give an example of a function h on $\mathbb{R}_+$ such that it is integrable and has the integrable T-transform but satisfies neither (1.90) nor (1.92), that is, belongs to H_o^1 but not to any of the embedded spaces in (1.93).

Before proceeding to the claimed example, we will calculate the T-transform of the indicator function $\chi_{[a,b]}(t)$ of the interval $[a,b]$ small enough and located far away from the origin; for a similar example for $\chi_{[0,b]}(t)$, see [61].

It is clear that $T\chi_{[a,b]}(x) = 0$ when $x < \frac{2a}{3}$ and $x > 2b$. Let now $x \in (a,b)$. We have

$$\int_{\frac{x}{2}}^{x-\delta} \frac{dt}{x-t} + \int_{x+\delta}^{\frac{3x}{2}} \frac{dt}{x-t} = \ln \left| \frac{x-a}{b-x} \right|,$$

and since it is independent of δ, this is exactly $T\chi_{[a,b]}$.

Further, $T\chi_{[a,b]}(x)$ takes the same value when either $x \in (\frac{2b}{3}, a)$ or $x \in (b, 2a)$. For $x \in (\frac{2a}{3}, \frac{2b}{3})$, we have

$$T\chi_{[a,b]}(x) = \int_a^{\frac{3x}{2}} \frac{dt}{x-t} = \ln \frac{2(a-x)}{x}.$$

1.4. Hardy spaces and subspaces

Similarly, when $x \in (2a, 2b)$

$$T\chi_{[a,b]}(x) = \int_{\frac{x}{2}}^{b} \frac{dt}{x-t} = \ln \frac{x}{2(x-b)}.$$

Let the function h we are going to construct be non-zero only on a family of intervals (a_k, b_k), where it is equal to A_k, with A_k being a monotone increasing positive sequence.

By this,

$$\int_{\mathbb{R}_+} |h(t)|\, dt = \sum_{k=1}^{\infty} A_k d_k < \infty, \tag{1.101}$$

where $d_k = b_k - a_k$.

Calculating the norm in O_p, we assume d_k to be small enough, while b_k is distant enough from b_{k-1}, at least so that only one (a_k, b_k) is located on each $(b_{k-1}, 2b_k)$. We then get

$$\sum_{k=2}^{\infty} \int_{b_{k-1}}^{b_k} \left(\frac{1}{x} \int_x^{2x} |h(t)|^p dt \right)^{\frac{1}{p}} dx \geq \sum_{k=2}^{\infty} \int_{b_{k-1}}^{b_k} \left(\frac{1}{x} \int_{a_k}^{b_k} |h(t)|^p dt \right)^{\frac{1}{p}} dx.$$

Routine calculations give that this value is not smaller than

$$\sum_{k=2}^{\infty} A_k (b_k - b_{k-1}) \left(\frac{d_k}{b_k} \right)^{\frac{1}{p}}. \tag{1.102}$$

It is clear that if $p = \infty$ we have the same value by formally considering $p = \infty$.

Let us now estimate the integral of $|Th|$. For this, we separately estimate

$$\int_{\frac{2a_k}{3}}^{2b_k} |Th(x)|\, dx.$$

As we have seen, there are 5 subintervals of $(\frac{2a_k}{3}, 2b_k)$ on which calculation of Th is slightly different. Integration of the results is similar over each of them; hence we consider one of the most problematic $(\frac{2b_k}{3}, a_k)$. The point is that, similarly to $(b_k, 2a_k)$, the length of the interval is not proportional to d_k. We get

$$\int_{\frac{2b_k}{3}}^{a_k} [\ln(a_k - x) - \ln(b_k - x)]\, dx$$

$$= d_k \ln \frac{1}{d_k (a_k - \frac{2b_k}{3})} + \frac{b_k}{3} \ln \left(1 - 3\frac{d_k}{b_k} \right).$$

Up to logarithmic factors, this is equivalent to d_k.

To conclude, we must have (1.101), even with possible logarithmic factors, but to diverge in O_p for each $p \in (1, \infty]$. For instance, let $d_k = 2^{-k}$, $b_k = 2^k$, and $A_k = k^{-\beta}2^k$. Here β should be large enough to ensure not only (1.101) but the same with possible logarithmic factors which are powers of k. For the sum in (1.102), we will get

$$\sum_{k=2}^{\infty} k^{-\beta} 2^{2(1-\frac{1}{p})k}.$$

Since $1 - \frac{1}{p} > 0$, the sum in (1.102) will be divergent.

By this we have the desired counterexample.

One more fact might be rather useful. There is a possibility to measure the norm in O_∞ in a different way.

By $L^* := L^*(\mathbb{R})$ we denote the space of functions f endowed with the norm

$$\|f\|_{L^*} = \int_0^\infty \operatorname*{ess\,sup}_{|t| \geq x} |f(t)|\, dx < \infty. \tag{1.103}$$

It differs slightly from the norm in O_∞ (first of all, f is a function on the whole real axis), but will be both convenient and used in the future.

Theorem 1.104. *We have $f \in L^*$ if and only if there exists a non-negative function F locally absolutely continuous on $(0, \infty)$ and such that almost everywhere $|f(\pm x)| \leq F(x)$ and $F'(x) \leq \lambda(x)$ for some positive function λ satisfying*

$$\int_0^\infty F(x)\, dx + \int_0^\infty x\lambda(x)\, dx < \infty.$$

To understand the properties of F and their relation to monotonicity, let us formulate and prove the following important fact for such functions F.

Lemma 1.105. *For a function F satisfying the assumptions of Theorem 1.104, we have*

$$\lim_{x \to \infty} xF(x) = 0. \tag{1.106}$$

This is by no means true for an arbitrary integrable function F, but the additional assumptions of the theorem weaker than the monotonicity guarantees (1.106). An analogous statement for sequences was proved in [28, Th.1]; the proof for functions goes along the same lines.

Proof. We have for $0 < x < y$

$$\int_x^y tF'(t)\, dt \leq \int_x^y t\delta(t)\, dt.$$

This is equivalent to the inequality

$$yF(y) - xF(x) - \int_x^y F(t)\, dt \leq \int_x^y t\lambda(t)\, dt,$$

1.4. Hardy spaces and subspaces

and hence
$$xF(x) - yF(y) \geq -\int_x^y F(t)\,dt - \int_x^y t\lambda(t)\,dt. \tag{1.107}$$

The right-hand side is negative and $o(1)$ as $x, y \to \infty$. Surely, $\varlimsup_{x\to\infty} xF(x)$ cannot be positive, otherwise $F(x)$ cannot be integrable. Hence there exists, for each positive ε, a sequence $\{x_n\}$ for which
$$x_n F(x_n) < \varepsilon.$$

Now suppose that $\varlimsup_{x\to\infty} xF(x) > 0$. Then there is a sequence $\{y_k\}$ such that
$$y_k F(y_k) > 2\varepsilon.$$

For $\{x_n\}$ and $\{y_{k_n}\}$, $y_{k_n} > x_n$, this contradicts (1.107), which yields (1.106). □

Proof of Theorem 1.104. Given $f \in L^*$, a natural desire is to take
$$\operatorname*{ess\,sup}_{|t|\geq x} |f(t)|$$

as $F(x)$. This function is monotone decreasing but may be not locally absolutely continuous. The following result due to I.A. Shevchuk (private communication) solves this problem and shows that F can be chosen to be pretty smooth.

Lemma 1.108. *For each monotone decreasing function g integrable on $(0, +\infty)$, arbitrary $\varepsilon > 0$, and any positive integer r, there is a monotone decreasing integrable function $G \in C^r(0, +\infty)$ such that*
$$G(x) \geq g(x) \tag{1.109}$$

for each $x \in (0, +\infty)$ and
$$\int_0^\infty [G(x) - g(x)]\,dx < \varepsilon. \tag{1.110}$$

Proof of Lemma 1.108. Since g is monotone decreasing and integrable on $(0, +\infty)$, there exists a sequence of points $\{x_k\}_{k=-\infty}^{+\infty}$ such that $x_k < x_{k-1}$ for all k, $\lim_{k\to+\infty} x_k = 0$ while $\lim_{k\to-\infty} x_k = +\infty$, and
$$\sum_{k=-\infty}^{+\infty} g(x_k)(x_{k-1} - x_k) - \int_0^\infty g(x)\,dx < \frac{\varepsilon}{2}. \tag{1.111}$$

Denote by $t_k \in (x_k, x_{k-1})$ the points satisfying $(t_k - x_k < x_k - x_{k+1})$
$$\sum_{k=-\infty}^{+\infty} [g(x_{k+1}) - g(x_k)](t_k - x_k) < \frac{\varepsilon}{2}. \tag{1.112}$$

For each k set
$$S_k(x) = \frac{\int_{x_k}^{x} (t_k - u)^r (u - x_k)^r \, du}{\int_{x_k}^{t_k} (t_k - u)^r (u - x_k)^r \, du}.$$

Finally, define G as
$$G(x) = \begin{cases} g(x_{k+1})(1 - S_k(x)) + g(x_k) S_k(x), & \text{if } x_k \leq x \leq t_k, \\ g(x_k), & \text{if } t_k < x \leq x_{k-1}. \end{cases}$$

Obviously, (1.109) holds, and
$$G(x) - g(x) = I_1(x) + I_2(x),$$
where $I_1(x) = g(x_k) - g(x)$ on each $[x_k, x_{k-1}]$, and
$$I_2(x) = \begin{cases} [g(x_{k+1}) - g(x_k)](1 - S_k(x)), & \text{if } x_k \leq x \leq t_k, \\ 0, & \text{if } t_k < x \leq x_{k-1}. \end{cases}$$

We obtain
$$\int_0^\infty I_2(x) \, dx = \sum_{k=-\infty}^{+\infty} (g(x_{k+1}) - g(x_k)) \int_{x_k}^{t_k} \frac{\int_x^{t_k} (t_k - u)^r (u - x_k)^r \, du}{\int_{x_k}^{t_k} (t_k - u)^r (u - x_k)^r \, du} \, dx,$$
and this value is bounded by the left-hand side of (1.112), while
$$\int_0^\infty I_1(x) \, dx$$
is exactly the left-hand side of (1.111). Hence we have (1.110), which completes the proof. □

Considering $g(x) = \operatorname*{ess\,sup}_{|t| \geq x} |f(t)|$, we may take the corresponding G for F. By this, we may take
$$\lambda(x) = -F'(x).$$
Then $\int_0^\infty F(x) \, dx < \infty$, and
$$\int_0^\infty F(x) \, dx = xF(x) \Big|_0^\infty - \int_0^\infty xF'(x) \, dx$$
$$= \int_0^\infty x\lambda(x) \, dx < \infty.$$

Let us prove the converse statement. Given a function F satisfying the assumptions of the theorem, let us show that $F \in L^*$. Since
$$\operatorname*{ess\,sup}_{|t| \geq x} |f(t)| \leq \operatorname*{ess\,sup}_{|t| \geq x} F(t) = \Phi(x),$$

1.4. Hardy spaces and subspaces

this will definitely prove the theorem. We have

$$\int_0^\infty \Phi(x)\,dx = \sum_{n=1}^\infty F(x_n)(x_n - a_n) + \sum_{n=1}^\infty \int_{x_{n-1}}^{a_n} F(x)\,dx,$$

where $x_0 = 0$, $a_n > x_{n-1}$ for $n > 1$, and $a_1 \geq x_0$. Here (x_{n-1}, a_n) are the intervals where $F(x)$ is monotone decreasing. Observe that $F(x_n) = F(a_n)$. We obtain

$$F(x_n)(x_n - a_n) = \int_{a_n}^{x_n} F(x)\,dx + \int_{a_n}^{x_n} xF'(x)\,dx,$$

and therefore

$$\int_0^\infty \Phi(x)\,dx = \int_0^\infty F(x)\,dx + \sum_{n=1}^\infty \int_{a_n}^{x_n} xF'(x)\,dx$$

$$\leq \int_0^\infty F(x)\,dx + \int_0^\infty x\lambda(x)\,dx < \infty. \tag{1.113}$$

Hence $\Phi \in L^*$, and the theorem is proved. □

For more details, relations with integrability of trigonometric series and the history of this specific result, see [111].

Now, we present a similar scale of nested spaces being subspaces of $H_e^1(\mathbb{R}_+)$. The construction comes from (1.59) and from the structure of O_q.

Definition 1.114. We define, for $1 < q \leq \infty$,

$$E_q := E_q(\mathbb{R}_+) := \left\{ g \in O_q : \int_0^\infty \frac{1}{x} \left| \int_0^x g(t)\,dt \right| dx < \infty \right\}.$$

The sum of the last integral and the norm in O_q gives the norm in E_q.

In addition to the above conditions which will be satisfied for the derivative f' in the results for Fourier transforms in the next chapters, the following one is also of notable interest and importance (see [137, 92]).

Definition 1.115. We say that a locally integrable function g defined on $\mathbb{R}_+$ belongs to $A_{1,2}$ if

$$\|g\|_{A_{1,2}} = \sum_{m=-\infty}^\infty \left\{ \sum_{j=1}^\infty \left[\int_{j2^m}^{(j+1)2^m} |g(t)|\,dt \right]^2 \right\}^{\frac{1}{2}} dx < \infty. \tag{1.116}$$

This space is amalgam in nature, since each of the summands in m is the norm in the Wiener amalgam space $W(L^1, \ell^2)$ for functions

$$G_m(t) = \begin{cases} 2^m g(2^m t), & if \quad t \geq 1, \\ 0, & otherwise, \end{cases}$$

where ℓ^p, $1 \leq p < \infty$, is a space of sequences $\{d_j\}$ endowed with the norm

$$\|\{d_j\}\|_{\ell^p} = \left(\sum_{j=-\infty}^{\infty} |d_j|^p\right)^{\frac{1}{p}}.$$

The norm of a function $h : \mathbb{R} \to \mathbb{C}$ in the amalgam space $W(L^1, \ell^2)$ is taken as (see, e.g., [55], [83])

$$\left\|\left\{\int_j^{j+1} |g(t)|\, dt\right\}\right\|_{\ell^2}.$$

In other words, we can rewrite (1.116) as

$$\|g\|_{A_{1,2}} = \sum_{m=-\infty}^{\infty} \|G_m\|_{W(L^1,\ell^2)} < \infty.$$

We have that $A_{1,2}$ is a subspace of L^1. Indeed, this follows from

$$\|g\|_{A_{1,2}} \geq \sum_{m=-\infty}^{\infty} \int_{2^m}^{2^{m+1}} |g(t)|\, dt = \|g\|_{L^1(\mathbf{R}_+)}. \tag{1.117}$$

For our aims, this can be reformulated as follows: if $f' \in A_{1,2}$, then f is of bounded variation, that is, $f' \in L^1(\mathbf{R}_+)$.

An example of a function in $A_{1,2}$ but not in $H_o^1(\mathbf{R}_+)$ will be given in Subsection 3.3.2 as well as the proof of the fact that the spaces $H_o^1(\mathbf{R}_+)$ and $A_{1,2}$ are incomparable. Without the latter statement, the effectiveness of the amalgam type results could be doubtful. All the O_q spaces are embedded in $A_{1,2}$ like they are embedded in $H_o^1(\mathbf{R}_+)$. Counterexamples for sequences can be found in [10] and [61].

1.4.4 A Paley–Wiener theorem

The argument in Example 1.73 and some other examples on the behavior of a function near infinity might be completely irrelevant in the case of odd functions. Indeed, by a Paley–Wiener theorem, if an odd integrable function is monotone decreasing (or general monotone in some sense, see [137] and [136]), no matter how slow, then its Hilbert transform is integrable [158]. This is not the case for even functions. The difference apparently comes from the fact that an odd function automatically has mean zero. This allows such an $H^1(\mathbb{R})$ function (more precisely, an $H_o^1(\mathbf{R}_+)$ function) to be of one sign on the half-axis.

The following matter is related to both this section and the previous one. It gives us a nice opportunity to again illustrate how the atomic characterization works and how effective it may be.

Let us first outline this problem in a more general context. It is well known that for the Hilbert transform $\mathcal{H}g(x)$ and the weight $w(x) = |x|^\alpha$ with $-1 < \alpha <$

1.4. Hardy spaces and subspaces

$p-1$, there holds $\|\mathcal{H}g\|_{L^p_w} \lesssim \|g\|_{L^p_w}$, $1 < p < \infty$. Here $L^p_w := L^p_w(\mathbb{R})$ means the weighted Lebesgue space endowed with the norm

$$\|g\|_{L^p_w} = \|g\|_{L^p_w(\mathbb{R})} = \left(\int_{\mathbb{R}} |g(t)|^p w(t)\,dt\right)^{\frac{1}{p}}, \qquad (1.118)$$

where the weight w is a non-negative locally integrable function. In [82], Hardy and Littlewood showed that for even functions g, this inequality also holds for $-p-1 < \alpha < p-1$. Later, Flett [57] proved the same results for odd functions provided $-1 < \alpha < 2p-1$. For $p=1$, it is known that only weak type inequality (1.55) holds for the Hilbert transform in general. On the other hand, Paley–Wiener's theorem asserts that for an odd and monotone decreasing on $\mathbb{R}_+$ function $g \in L^1$ one has $\mathcal{H}g \in L^1$. In [137], this theorem was extended to general monotone functions. Further, in [136] the weighted analogues of the Paley–Wiener theorem for odd and even (general monotone) functions were proved. In other words, it was an extension of Hardy–Littlewood's [82], Flett's [57] and Andersen's [8] results to the case $p=1$ under the assumption of (general) monotonicity for an even/odd function.

Besides the initial proof in [158] (for series) and additional study in [210], a different proof of the initial Paley–Wiener theorem can be found in [173, Ch.IV, 6.2]. Let us give details. First of all, integrability of the Hilbert transform of an integrable function means that this function belongs to the real Hardy space $H^1(\mathbb{R})$.

Let g_0 be a non-negative monotone decreasing function on $(0, \infty)$ such that

$$\int_0^{\infty} g_0(t)\,dt < \infty,$$

and let $g(t) = g_0(t)$ on $(0, \infty)$, and $g(-t) = -g(t)$. Therefore, the Paley–Wiener theorem then states that $g \in H^1(\mathbb{R})$. The proof in [173, Ch.IV, 6.2] goes along the following lines. For $-\infty < k < \infty$, let

$$a_k(t) = \frac{1}{2^{k+2}g_0(2^k)} g_0(|t|)\operatorname{sign} t$$

if $2^k \leq |t| < 2^{k+1}$ and zero otherwise. Obviously, each a_k is an atom (in fact, a $(1, \infty, 0)$ atom). To see that the absolute value of the function is less than the reciprocal of the length of the support interval $[-2^{k+1}, 2^{k+1}]$, just the monotonicity is used. Taking $\lambda_k = 2^{k+2}g_0(2^k)$ and observing that $\sum_{k=-\infty}^{\infty} \lambda_k \leq 8\int_0^{\infty} g_0(t)\,dt$, we see that the series $\sum_{k=-\infty}^{\infty} \lambda_k a_k(t)$ converges to $g(t)$ except at the origin. Since thus we have an atomic decomposition of g, it belongs to $H^1(\mathbb{R})$.

We can immediately extend both this result and its proof by taking g to be weak monotone. To define the latter notion, we will assume a function to lie on $(0, \infty)$, to be locally of bounded variation, and vanishing at infinity.

Definition 1.119. We say that a non-negative function f defined on $(0, \infty)$, is *weak monotone*, written WM, if

$$f(t) \leq Cf(x) \quad \text{for any } t \in [x, 2x]. \tag{1.120}$$

Using in the above proof $g_0(t) \leq Cg_0(2^k)$ provided $g_0 \in WM$ instead of $g_0(t) \leq g_0(2^k)$ for monotone g_0, we immediately arrive at the following more general result than Theorem 6.1 in [137], since the notion of weak monotonicity introduced and widely used in [138] is less restrictive than certain notions of general monotonicity (see [137]).

Theorem 1.121. *Let g_0 be integrable on $(0, \infty)$ and $g_0 \in WM$. Then $g \in H^1(\mathbb{R})$, or, equivalently, its Hilbert transform is integrable.*

This shows that for the integrability of the Hilbert transform smoothness conditions are frequently not of crucial importance; certain regularity of the functions works instead.

1.5 Balance integral operator

We define the *balance integral operator* with kernel φ by its action on an appropriate function $g : \mathbb{R}_+ \to \mathbb{C}$ as

$$B_\varphi g(x) = \frac{1}{x^2} \int_0^\infty g\left(\frac{t}{x}\right) \varphi(t) \, dt. \tag{1.122}$$

Of course, in a somewhat artificial way (first of all by considering its action on functions $G(t) = g(\frac{1}{t})$), this operator can be related to a certain Hausdorff type operator (see, e.g., [116]) or to some multiplicative convolution, but, for brevity and convenience, we prefer to keep its form and name as it is. Its balance role will become clear later. In fact, it will seem to be extraneous till Chapter 4, where its rightful place in the asymptotic formulas for the Fourier transforms will be revealed. However, to be precise, one of its versions, B_s, with $\varphi(\cdot) = \sin(\cdot)$, that is,

$$\frac{\widehat{g_s}(x)}{x} = B_s g(x)$$

will appear already in Chapter 3. It is not our goal to study this operator in full detail, though this might be rather useful and important.

Similarly to the above, $B_c g(x)$, with $\varphi(\cdot) = \cos(\cdot)$, is the cosine Fourier transform of g times x. For the general Fourier transform, one may consider B_φ with $\varphi(\cdot) = e^{i\cdot} + e^{-i\cdot}$. In these forms, the oscillating nature of the Fourier transforms is partially hidden in the function itself. Further, for odd functions, (1.61) can be rewritten as

$$\int_0^\infty |B_s g(x)| \, dx \lesssim \|g\|_{H^1(\mathbb{R})}. \tag{1.123}$$

1.5. Balance integral operator

The latter also shows that the importance and applications of B_φ strongly depend on the generating function φ and on the spaces on which the operator is acting.

By $L_w^p(\mathbb{R}_+)$ we denote the weighted Lebesgue space, with the same norm as in (1.118) but with integration over $\mathbb{R}_+$ rather than over $\mathbb{R}$. Of course, they become usual $L^p(\mathbb{R}_+)$ or $L^p(\mathbb{R})$ is the case where the weight is constant.

Lemma 1.124. *For $g \in L_w^p(\mathbb{R}_+)$, $1 \le p < \infty$, with the weight $w(t) = t^{2p-2}$, we have $B_\varphi g \in L^p(\mathbb{R}_+)$ provided*

$$\int_0^\infty |\varphi(t)| t^{\frac{1}{p}-2}\, dt < \infty.$$

Proof. This is just Minkowski's inequality and simple substitutions, which lead to the inequality

$$\left(\int_0^\infty |B_\varphi g(x)|^p\, dx\right)^{\frac{1}{p}} \le \int_0^\infty |\varphi(t)| t^{\frac{1}{p}-2}\, dt \left(\int_0^\infty x^{2p-2} |g(x)|^p\, dx\right)^{\frac{1}{p}}. \quad (1.125)$$

Here the condition on φ may also be written as $\varphi \in L_v^1(\mathbb{R}_+)$, with the weight $v(t) = t^{\frac{1}{p}-2}$. $\square$

By letting $p = 1$ in (1.125), we obtain a useful corollary.

Corollary 1.126. *For $g \in L^1(\mathbb{R}_+)$, we have $B_\varphi \in L^1(\mathbb{R}_+)$ provided*

$$\int_0^\infty \frac{|\varphi(t)|}{t}\, dt < \infty. \quad (1.127)$$

Of course, similar estimates can be obtained for a variety of weighted spaces by applying more advanced tools. However, even the above simple machinery shows the following fact concerning the boundedness of balance operators in the above considered important scale of spaces.

Lemma 1.128. *For $g \in O_q$, $1 < q \le \infty$, we have $B_\varphi g \in O_q$ provided (1.127) holds.*

Proof. Instead of estimating

$$\int_0^\infty \left(\frac{1}{x} \int_x^{2x} \left| \frac{1}{u^2} \int_0^\infty g\left(\frac{t}{u}\right) \varphi(t)\, dt \right|^q du\right)^{\frac{1}{q}} dx,$$

we can estimate the equivalent quantity

$$\int_0^\infty x^{-2-\frac{1}{q}} \left(\int_x^{2x} \left| \int_0^\infty g\left(\frac{t}{u}\right) \varphi(t)\, dt \right|^q du\right)^{\frac{1}{q}} dx.$$

Applying the generalized Minkowski inequality to the inner integrals, we arrive at the bound

$$\int_0^\infty x^{-2-\frac{1}{q}} \int_0^\infty |\varphi(t)| \left(\int_x^{2x} \left| g\left(\frac{t}{u}\right) \right|^q du\right)^{\frac{1}{q}} dt\, dx.$$

Substituting $\frac{t}{u} \to u$, we get

$$\int_0^\infty x^{-2-\frac{1}{q}} \int_0^\infty t^{-\frac{1}{q}} |\varphi(t)| \left(\int_{\frac{t}{2x}}^{\frac{t}{x}} |g(u)|^q du \right)^{\frac{1}{q}} dt\, dx$$

$$= \int_0^\infty t^{-\frac{1}{q}} |\varphi(t)| \int_0^\infty x^{-2-\frac{1}{q}} \left(\int_{\frac{t}{2x}}^{\frac{t}{x}} |g(u)|^q du \right)^{\frac{1}{q}} dx\, dt.$$

Substituting now $\frac{t}{2x} \to x$, we derive that the right-hand side is dominated by $\|g\|_{O_q}$ times the integral in condition (1.127), which completes the proof for $q < \infty$.

If $q = \infty$, appropriate changes should be taken. We estimate

$$\int_0^\infty \operatorname*{ess\,sup}_{u \le x \le 2u} \frac{1}{u^2} \left| \int_0^\infty g\left(\frac{t}{u}\right) \varphi(t)\, dt \right| dx$$

$$\le \int_0^\infty \int_0^\infty \operatorname*{ess\,sup}_{u \le x \le 2u} \frac{1}{u^2} \left| g\left(\frac{t}{u}\right) \right| |\varphi(t)|\, dt\, dx.$$

By Fubini's theorem, this is equal to

$$\int_0^\infty |\varphi(t)| \int_0^\infty \operatorname*{ess\,sup}_{u \le x \le 2u} \frac{1}{u^2} \left| g\left(\frac{t}{u}\right) \right| dx\, dt.$$

Changing variables $\frac{t}{u} = v$, we get the equal value

$$\int_0^\infty |\varphi(t)| \int_0^\infty \frac{1}{x^2} \operatorname*{ess\,sup}_{\frac{t}{v} \le x \le \frac{2t}{v}} |g(v)|\, dx\, dt.$$

Changing variables once more, $\frac{t}{x} = z$, say, we see that the last quantity is dominated by $\|g\|_{O_\infty}$ times the integral in (1.126), as required. □

Of course, such an operator also makes sense if defined not on $\mathbb{R}_+$ but on $\mathbb{R}$ or on a finite interval $[a, b]$. The definition should be adjusted in an appropriate way, which should not lead to any substantial difficulties. In fact, the considered above operator is a partial case (with $a = 0$ and $b = \infty$) of the more general one defined on arbitrary $[a, b]$ by

$$B_\varphi g(x) := B_{\varphi(\cdot, x)} g(x) = \frac{1}{x^2} \int_{ax}^{bx} g\left(\frac{t}{x}\right) \varphi(t, x)\, dt. \tag{1.129}$$

When the latter version is used, we shall keep the previous notation B_φ without indicating the dependence on a and b and more complicated structure of the kernel φ, since this will not result in any confusion.

By association of definition and ideas, recall the notion of the weighted Hardy–Littlewood average operator. It is introduced in [38] as

$$U_\varphi g(x) = \int_0^1 g(tx) \varphi(t)\, dt.$$

1.5. Balance integral operator

Here x is assumed not only to be a real number but mainly $x \in \mathbb{R}^n$. It is not difficult to understand that for a function φ supported on $[0, 1]$,

$$B_\varphi g(x) = \frac{1}{x^2} U_\varphi g\left(\frac{1}{x}\right).$$

Except the initial [38], there are other works on properties and estimates for U_φ. We mention only one of them [207].

Chapter 2

Functions with derivative in a Hardy space

As explained above, we are going to study separately the cosine Fourier transform

$$\widehat{f_c}(x) = \int_0^\infty f(t) \cos xt \, dt$$

and the sine Fourier transform

$$\widehat{f_s}(x) = \int_0^\infty f(t) \sin xt \, dt,$$

and their integrability properties. The otherness of the two transforms is a known phenomenon. For instance, if a function is monotone on $\mathbb{R}_+ = [0, \infty)$, then its sine Fourier transform preserves the sign and furnishes more information (see, e.g., Section 2.1 and [135]). On the other hand, the Fourier transform of such a function cannot be Lebesgue integrable on $\mathbb{R}$. For a convex function, the classical Pólya theorem (see, e.g., Corollary 2.30 and [141], [186, 6.10, Thm. 124]) brings the cosine transform to the fore. More precisely, the cosine Fourier transform of a bounded, continuous, convex function, which vanishes at infinity, preserves the sign and is Lebesgue integrable no matter how slow the function decays near infinity.

The other issue is as follows. While studying conditions for integrability of the Fourier transform, those for the integrability near infinity are usually a point of interest. This is frequently of crucial importance in applications (see detailed analysis of this in [92]). One of the reasons is that for Lebesgue integrable functions integrability of the Fourier transform on a compact set is an obvious fact. But since we here deal with possibly non-integrable functions, integrability near the origin is of interest and equal importance as well. For instance, the example of an even monotone function with non-integrable Fourier transform given in [186, 6.11, Theorem 125] is the one where the Fourier transform is not integrable on $[0,1]$. The reader can find many important results on this and related issues in more recent books [36, 198, 92].

What is special about these problems is the case of a function with bounded variation. A series of works has appeared during last 25 years where the Fourier transform of a function of bounded variation is studied; see, e.g., [198], [68], [107], [61], [135], [92], etc. It may be added that Trigub's results on the asymptotic behavior of the Fourier transform of a convex function (see, e.g., [191], [198] or Theorem 4.3 in Chapter 4 of this book; for their improvement to convex functions with singularities, see [107] and [92, Ch.3]) are the reference point for the consequent activity. Recall that we deal with functions f of bounded variation on $\mathbb{R}_+ = [0,\infty)$, which vanish at infinity, $\lim_{t\to\infty} f(t) = 0$, written $f \in BV_0[0,\infty)$, and locally absolutely continuous on $(0,\infty)$, written $f \in LAC(0,\infty)$. We will denote these conditions by

$$f \in BV_0[0,\infty) \cap LAC(0,\infty).$$

However, before proceeding to exactly such functions, let us consider certain examples and, so to say, warm-up results from [135]. After that, we provide a systematic study of the Fourier transforms of functions with derivative in a certain Hardy space (see mainly [123] and [126]). In the end, we discuss certain issues related to absolute continuity.

2.1 First steps

Let us start with a few examples. First, if $f(t)$ is $t^{-\frac{1}{2}}$, then its Fourier transforms $\widehat{f}_c$ and $\widehat{f}_s$ are identical and equal to

$$\widehat{f}_c(x) = \widehat{f}_s(x) = \sqrt{\frac{\pi}{2x}}.$$

They are integrable over $(0,1)$ but not near infinity. Then, if $f(t)$ is e^{-t}, we have

$$\widehat{f}_c(x) = \frac{1}{1+x^2}$$

and

$$\widehat{f}_s(x) = \frac{x}{1+x^2}.$$

In this case, $\widehat{f}_c$ is integrable over $(0,+\infty)$ while $\widehat{f}_s$ is integrable only on finite intervals. Finally, if $f(t)$ is t^{-1}, we obtain $\widehat{f}_s(x) = \frac{\pi}{2}$ for all $x > 0$ while $\widehat{f}_c(x)$ just does not exist as an improper integral. These examples give us a general idea of what to expect near infinity and near the origin and can be found in [15].

The results we give below are probably a sort of folklore but have never appeared till recently in an accurate form, see [135]. Clear hints are given by the theory of trigonometric series with monotone coefficients. For applications, we

2.1. First steps

are interested in bounded functions but some of our results are valid for functions infinitely growing near the origin. Our functions may be not integrable on the whole half-axis $\mathbb{R}_+ = [0, \infty)$, hence the integrals are understood as improper integrals.

Theorem 2.1. *For f locally absolutely continuous on $(0, +\infty)$, vanishing at infinity $\lim_{t \to \infty} f(t) = 0$, and monotone,*

$$\int_0^\pi |\widehat{f_c}(x)|\, dx \leq \pi \int_0^1 |f(t)|\, dt + 3 \int_1^\infty \frac{|f(t)|}{t}\, dt \qquad (2.2)$$

and

$$\int_0^1 t|f(t)|\, dt + \frac{1}{12} \int_1^\infty \frac{|f(t)|}{t}\, dt$$
$$\leq \int_0^\pi |\widehat{f_s}(x)|\, dx \qquad (2.3)$$
$$\leq \frac{\pi^2}{2} \int_0^1 t|g(t)|\, dt + 2 \int_1^\infty \frac{|f(t)|}{t}\, dt.$$

Proof. Let us begin with the cosine transform. We have

$$\int_0^\infty f(t) \cos xt\, dt = \int_0^{\frac{\pi}{x}} f(t) \cos xt\, dt - \frac{1}{x} \int_{\frac{\pi}{x}}^\infty f'(t) \sin xt\, dt. \qquad (2.4)$$

Integrating the first integral on the right modulo, we obtain

$$\int_0^\pi \left| \int_0^{\frac{\pi}{x}} f(t) \cos xt\, dt \right| dx \leq \int_0^1 |f(t)| \int_0^\pi |\cos xt|\, dx\, dt$$
$$+ \int_1^\infty |f(t)| \int_0^{\frac{\pi}{t}} |\cos xt|\, dx\, dt \leq \pi \int_0^1 |f(t)|\, dt$$
$$+ 2 \int_1^\infty \frac{|f(t)|}{t}\, dt. \qquad (2.5)$$

In the second integral on the right-hand side of (2.4), we merely use the monotonicity of f and rough estimates. By this we arrive at (2.2).

It is clear that among the above examples, only $f(t) = t^{-1}$ does not satisfy (2.2) in the sense that the right-hand side of (2.2) is not finite for it.

For the sine transform, $\widehat{f_s}$ is of the same sign as f is. This can be found in, e.g., [186, 6.10, Theorem 123]; however, let us prove this nice property here. Let, for simplicity, f be monotone decreasing, that is, positive. Given $x > 0$, we have

$$\int_0^\infty f(t) \sin xt\, dt = \sum_{k=0}^\infty \left[\int_{\frac{2k\pi}{x}}^{\frac{(2k+1)\pi}{x}} + \int_{\frac{(2k+1)\pi}{x}}^{\frac{(2k+2)\pi}{x}} \right] f(t) \sin xt\, dt.$$

Substituting $t \to t + \frac{\pi}{x}$ in the second integral on the right, we obtain

$$\int_0^\infty f(t) \sin xt\, dt = \sum_{k=0}^\infty \int_{\frac{2k\pi}{x}}^{\frac{(2k+1)\pi}{x}} \left[f(t) - f\left(t + \frac{\pi}{x}\right) \right] \sin xt\, dt.$$

The difference in the brackets is positive due to the monotonicity of f, also $\sin xt$ is positive on $\left(\frac{2k\pi}{x}, \frac{(2k+1)\pi}{x}\right)$, and we are done.

Again, let, for simplicity, f be monotone decreasing with, consequently, non-negative Fourier transform. We have

$$\int_0^\pi |\widehat{f}_s(x)|\, dx = \int_0^\pi \int_0^\infty f(t) \sin xt\, dt\, dx$$
$$= \int_0^2 f(t)\, dt \int_0^\pi \sin xt\, dt + \int_2^\infty f(t)\, dt \int_0^{\frac{2\pi}{t}} \sin xt\, dt$$
$$+ \int_2^\infty f(t)\, dt \int_{\frac{2\pi}{t}}^\pi \sin xt\, dt$$
$$= 2 \int_0^\infty \frac{f(t)}{t} \sin^2 \frac{\pi t}{2}\, dt. \tag{2.6}$$

The right-hand side of (2.6) is

$$2\left(\int_0^1 + \int_1^\infty\right) \frac{f(t)}{t} \sin^2 \frac{\pi t}{2}\, dt$$
$$\leq \frac{\pi^2}{2} \int_0^1 tg(t)\, dt + 2 \int_1^\infty \frac{f(t)}{t}\, dt.$$

The estimate from below is derived as follows:

$$\int_0^\pi |\widehat{f}_s(x)|\, dx \geq 2 \int_0^1 \frac{f(t)}{t} \sin^2 \frac{\pi t}{2}\, dt + 2 \int_{\frac{5}{2}}^\infty \frac{f(t)}{t} \sin^2 \frac{\pi t}{2}\, dt$$
$$\geq 2 \int_0^1 tf(t)\, dt + 2 \sum_{k=1}^\infty \int_{2k+\frac{1}{2}}^{2k+1} \frac{f(t)}{t} \sin^2 \frac{\pi t}{2}\, dt.$$

To estimate the sum on the right, we observe that on each $(2k + \frac{1}{2}, 2k + 1)$ there holds

$$\sin^2 \frac{\pi t}{2} = \sin^2 \frac{(t-2k)\pi}{2} \geq \frac{1}{2}$$

and, by the monotonicity of $t^{-1}f(t)$,

$$3 \int_{2k+\frac{1}{2}}^{2k+1} \frac{f(t)}{t}\, dt \geq \int_{2k+1}^{2k+2+\frac{1}{2}} \frac{f(t)}{t}\, dt. \tag{2.7}$$

We get

$$\int_0^\pi |\widehat{f_s}(x)|\,dx \geq 2\int_0^1 tf(t)\,dt + \frac{1}{6}\int_{5/2}^\infty \frac{f(t)}{t}\,dt.$$

Next, using again (2.7) with $k=0$, we obtain

$$\int_0^1 tf(t)\,dt \geq \int_{\frac12}^1 t^2 \frac{f(t)}{t}\,dt \geq \frac{1}{12}\int_1^{\frac52} \frac{f(t)}{t}\,dt.$$

This is the lower bound in (2.3), and the proof is complete. $\square$

We have considered integrability conditions only on $[0,\pi]$; similarly, one may study the case $x \in (\pi,\infty)$. However, since the above results are given only as an introduction to the main ones, we omit that counterpart.

In [135], the classes of so-called general monotone functions are introduced as an analog of the same notion for sequences earlier introduced by Tikhonov in [185] (for a comprehensive survey of these classes, see [137]). The study of the Fourier transform for such classes is closely related to the study of the Fourier transforms for functions of bounded variation.

2.2 Derivative in $H_o^1(\mathbb{R}_+)$

This section is devoted to the theorem that is the main result of [92, Ch.3]. It was proven first in [107]; a different proof, with the help of a (somewhat special) atomic decomposition, was later given in [61]. Is there any motive to return to this issue? Well, the reason is more than sufficient. We will give a completely different proof of it (it has recently appeared in [123]). This has not only intrinsic interest but also mainly because the idea of the proof gives rise to completely new results for different classes, and, more than this, to sharper relations for the Fourier transform of a function of bounded variation; see the next sections.

Theorem 2.8. *Let $f \in BV_0[0,\infty) \cap LAC(0,\infty)$ and $f' \in H_o^1(\mathbb{R}_+)$. Then the cosine Fourier transform of f is integrable, with*

$$\|\widehat{f_c}\|_{L^1(\mathbb{R}_+)} \lesssim \|f'\|_{H_o^1(\mathbb{R}_+)}; \tag{2.9}$$

while for the sine Fourier transform an asymptotic formula holds: for $x > 0$,

$$\widehat{f_s}(x) = \frac{1}{x}f\left(\frac{\pi}{2x}\right) + F(x), \tag{2.10}$$

where

$$\|F\|_{L^1(\mathbb{R}_+)} \lesssim \|f'\|_{H_o^1(\mathbb{R}_+)}. \tag{2.11}$$

Proof. The first step – integration by parts – is standard and inevitable. In fact, this reflects one of the main features of a function of bounded variation: such a function is almost everywhere differentiable and the derivative is Lebesgue integrable. The latter ensures that after integration by parts the Fourier transform of the derivative will be well defined. By this, (2.9) immediately follows from (1.61) applied to the $H_o^1(\mathbb{R}_+)$. Let us proceed to the more disputable issue, to the sine transform. After integration by parts, instead of the sine transform we must estimate

$$\frac{f(0)}{x} + \frac{1}{x}\int_0^\infty f'(t)\cos xt\, dt. \tag{2.12}$$

Needless to say, $f(0)$ is understood as the right limit. Now we wish to use the inversion formula (e.g., see [100, 1, (4.18) or (5.46)]):

$$\frac{1}{x}\int_0^\infty f'(t)\cos xt\, dt = -\frac{1}{x}\int_0^\infty \mathcal{H}_o(\mathcal{H}_e f')(t)\cos xt\, dt. \tag{2.13}$$

Since we assume that, for f' only the odd Hilbert transform is integrable, we must justify this order of application of the Hilbert transforms. Obviously (cf. Proposition 1.58), for an integrable function g,

$$\mathcal{H}_e g(u) = \mathcal{H}_o g(u) + \frac{2}{\pi}\int_0^\infty \frac{g(s)}{s+u}\, dt \tag{2.14}$$

for any integrable function g. By representing the last integral in the form

$$\int_0^\infty \frac{g(s)}{s+u}\, dt = \frac{1}{u}\int_0^u g(s)\, ds$$
$$- \int_0^u g(s)\frac{t}{u(s+u)}\, ds + \int_u^\infty \frac{g(s)}{s+u}\, ds, \tag{2.15}$$

we readily see that the last two terms on the right-hand side are integrable. The integrability of the first term on the right-hand side in (2.14) follows from the assumption of the theorem. The odd Hilbert transform can be applied to the first term on the right-hand side in (2.15) as follows. Denoting

$$\Phi(u) = \frac{1}{u}\int_0^u g(s)\, ds,$$

we represent $\Phi(u) = \Phi_1(u) + \Phi_2(u)$, where

$$\Phi_1(u) = \begin{cases} \frac{1}{u}\int_0^u g(s)\, ds, & u \geq 1, \\ 0, & \text{otherwise,} \end{cases}$$

and

$$\Phi_2(u) = \begin{cases} \frac{1}{u}\int_0^u g(s)\, ds, & u \leq 1, \\ 0, & \text{otherwise.} \end{cases}$$

2.2. Derivative in $H_o^1(\mathbb{R}_+)$

Since $\Phi_1 \in L^2$, its Hilbert transform is well-defined. Representing $\frac{1}{t^2-u^2}$ as

$$\frac{1}{2t}\left[\frac{1}{u+t} - \frac{1}{u-t}\right]$$

and seeing that the first term in the brackets does not cause any problems, we arrive at the Hilbert transform times a constant of $u\Phi_2(u)$. But this function belongs to any L^p and thus its Hilbert transform is well-defined and exists almost everywhere.

Further, by transposing the Hilbert transform from $\mathcal{H}_e f'$ to $\cos xt$ on the right-hand side in (2.13), we obtain

$$-\frac{1}{x}\int_0^\infty \frac{2}{\pi}\int_0^\infty \frac{u(\mathcal{H}_e f')(u)}{t^2-u^2}\,du\,\cos xt\,dt$$

$$= \frac{1}{x}\int_0^\infty (\mathcal{H}_e f')(u)\left[\frac{2u}{\pi}\int_0^\infty \frac{\cos xt}{u^2-t^2}\,dt\right]du$$

$$= \frac{1}{x}\int_0^\infty (\mathcal{H}_e f')(u)(\mathcal{H}_e \cos(x\cdot))(u)\,du.$$

The Hilbert transform of $\cos xt$ is $\sin xt$ (see Example 1.26 or, e.g., [100, Vol 2, Table 1.5 (5.2)]), which transforms (2.12) into

$$\frac{f(0)}{x} + \frac{1}{x}\int_0^\infty (\mathcal{H}_e f')(t)\sin xt\,dt. \tag{2.16}$$

We cannot apply the Fourier-Hardy inequality in the current context, because the last integral contains $\mathcal{H}_e$ rather than $\mathcal{H}_o$. Of course, the first term prevents us from obtaining straightforward estimates as well. Let us carry out appropriate transformations.

By substituting the integral on the right-hand side in (2.14) (with $g = f'$) into the sine transform on the right-hand side in (2.23), we see that we must estimate the expression

$$\frac{2}{\pi x}\int_0^\infty \int_0^\infty \frac{f'(u)}{t+u}\,du\,\sin xt\,dt = \frac{2}{\pi x}\int_0^\infty f'(u)\int_0^\infty \frac{\sin xt}{t+u}\,dt\,du. \tag{2.17}$$

Note that plain integration by parts gives

$$\int_0^\infty \frac{\sin xt}{t+u}\,dt = O\left(\frac{1}{xu}\right), \tag{2.18}$$

and hence we readily see that if we take the outer integral on the right-hand side in (2.17) over the half-interval $[\frac{\pi}{2x}, \infty)$ and apply (2.18), then the integral can be estimated as

$$\frac{2}{\pi}\int_0^\infty \frac{1}{x^2}\int_{\frac{\pi}{2x}}^\infty \frac{|f'(u)|}{u}\,du\,dx = \frac{2}{\pi}\int_0^\infty \frac{|f'(u)|}{u}\,du\int_{\frac{\pi}{2x}}^\infty \frac{1}{x^2}\,dx$$

$$= \frac{4}{\pi^2}\int_0^\infty |f'(u)|\,du. \tag{2.19}$$

Using the well-known relation
$$\int_0^\infty \frac{\sin xt}{t}\,dt = \frac{\pi}{2},$$
we obtain
$$\frac{2}{\pi x}\int_0^{\frac{\pi}{2x}} f'(u)\int_0^\infty \frac{\sin xt}{t}\,dt\,du = \frac{1}{x}f\left(\frac{\pi}{2x}\right) - \frac{f(0)}{x},$$
which not only gives the leading term in (2.10) but also eliminates the problematic first term in (2.23). It remains to estimate
$$\int_0^\infty \frac{1}{x}\left|\int_0^{\frac{\pi}{2x}} uf'(u)\int_0^u \frac{\sin xt}{t(u+t)}\,dt\,du\right|dx.$$

To this end, we split the inner integral into integrals over $[0,u]$ and (u,∞) and treat the two integrals separately. In the first case, we replace $|\sin xt|$ by xt and note that
$$\int_0^u \frac{dt}{u+t} = \ln 2,$$
which leads to the estimate
$$\int_0^\infty \int_0^{\frac{\pi}{2x}} u|f'(u)|\,du\,dx = \frac{\pi}{2}\int_0^\infty |f'(u)|\,du.$$

In the second case, we replace $|\sin xt|$ by $\sqrt{xt}$, use the fact that
$$\int_u^\infty \frac{dt}{\sqrt{t}(u+t)} = O\left(\frac{1}{\sqrt{u}}\right),$$
and complete the calculations with the estimate
$$\int_0^\infty \frac{1}{\sqrt{x}}\int_0^{\frac{\pi}{2x}} \sqrt{u}|f'(u)|\,du\,dx = \sqrt{2\pi}\int_0^\infty |f'(u)|\,du.$$

Let us give a preliminary summary of the results. We have obtained the relation
$$\widehat{f}_s(x) = \frac{1}{x}f\left(\frac{\pi}{2x}\right) + \frac{1}{x}\int_0^\infty (\mathcal{H}_o f')(t)\sin xt\,dt + \Phi(x),$$
where
$$\int_0^\infty |\Phi(x)|\,dx = O\left(\int_0^\infty |f'(x)|\,dx\right).$$

This relation is, undoubtedly, of interest in itself. It is primarily important to us in that it readily opens the way to completing the proof. By taking $\mathcal{H}_o f'$ for g in (1.61) and by using the inversion formula for the Hilbert transform, we obtain the desired estimate (2.11). □

2.3. Derivative in $H^1_e(\mathbb{R}_+)$

It is important to note the following when analyzing this proof and recalling the earlier proofs. The estimates in the cited papers have not actually been carried out for the odd Hilbert transform but rather for the transform (1.94).

The difference in the action of these two operators on an integrable odd function is always an integrable function, and so the two transforms are equivalent in the problems in question. Nevertheless, it is worth noting that our proof does not involve any auxiliary transforms; we never move beyond the framework of the Hilbert transform and the Hardy space. Further, it turns out that the well-known difference between the cosine and sine transforms, expressed by the leading term $\frac{1}{x} f\left(\frac{\pi}{2x}\right)$ in the asymptotic relation (2.10), is essentially the difference between the even and odd Hilbert transforms. All these show that (1.94) is a somewhat artificial substitute for the odd Hilbert transform in our considerations. However, it might have a certain convenience in calculations (see, e.g., Subsection 1.4.3 in Chapter 1 or [113]).

2.3 Derivative in $H^1_e(\mathbb{R}_+)$

The reader can easily observe a certain lopsidedness in the above results and those in the next section: the fact that the derivative of such a function belongs to each of the classes studied before ensures the integrability of the *cosine* Fourier transform, while the *sine* Fourier transform of a function from exactly the same class sometimes enjoys an asymptotic relation, with the leading term $\frac{1}{x} f\left(\frac{\pi}{2x}\right)$ and an integrable remainder, whose L^1 norm is controlled by the norm of the derivative in the considered space. By this, integrability or non-integrability of the sine Fourier transform depends on the integrability or non-integrability of that leading term. Till recently this was the case in all results of the considered type.

This phenomenon is not new for both Fourier series and Fourier transforms, see, e.g., Section 2.1. However, one may guess that there should exist a setting where the cosine Fourier transform is subject to a more delicate analysis. The main result of this section shows that such a setting indeed exists, and also within the framework of the theory of Hardy spaces. The latter is important, since beyond that scope the sine Fourier transform comes in the vanguard, as shown in [121] and in the next section.

Now, in terms of assumptions our main result is an obvious counterpart of Theorem 2.8.

Theorem 2.20. *Let $f \in BV_0[0, \infty) \cap LAC(0, \infty)$ and $f' \in H^1_e(\mathbb{R}_+)$. Then the sine Fourier transform of f is integrable, with*

$$\|\widehat{f_s}\|_{L^1(\mathbb{R}_+)} \lesssim \|f'\|_{H^1_e(\mathbb{R}_+)};$$

while for the cosine Fourier transform an asymptotic formula holds: for $x > 0$,

$$\widehat{f_c}(x) = \frac{A}{x} f\left(\frac{\pi}{2x}\right) + \frac{2}{\pi x} \int_0^{\frac{\pi}{2x}} f'(u) \ln \frac{2ux}{\pi} \, du + F(x), \qquad (2.21)$$

where
$$A = \frac{2}{\pi} \left(\int_0^{\frac{\pi}{2}} \frac{1-\cos t}{t} \, dt - \int_{\frac{\pi}{2}}^\infty \frac{\cos t}{t} \, dt \right)$$
and
$$\|F\|_{L^1(\mathbb{R}_+)} \lesssim \|f'\|_{H_e^1(\mathbb{R}_+)}.$$

Remark 2.22. Unlike Theorem 2.8, one can see here an additional leading term. If $\frac{f(u)}{u}$ is integrable near zero, integration by parts makes the second leading term in (2.21) to be
$$\frac{1}{x} \int_0^{\frac{\pi}{2x}} \frac{f(u)}{u} \, du.$$

The possibility to get this result is not only the matter of a function class but also of a proof (cf. [123] and the previous section). The proofs known earlier led only to Theorem 2.8 or similar ones. In other words, certain asymmetry always appears, even in situations that seem completely symmetric. One of the reasons may lie in the fact that odd and even functions admit the mean zero property – absolutely necessary for functions in the real Hardy space – in a different manner: the former possesses it automatically while the latter needs to satisfy it on the half-axis for having it on the whole axis, see (1.80). Because of this or for other reasons, previous proofs were insensitive to the space $H_e^1(\mathbb{R}_+)$. Indeed, looking through the proof of Theorem 2.8 in [107] or [92, Ch.3], one sees that the space $H_o^1(\mathbb{R}_+)$ appears there on its own account, while in the proof in the previous section it was put into effect "by force". In the same way, $H_e^1(\mathbb{R}_+)$ will appear in the following proof.

Proof. Integrating by parts, we obtain for the sine Fourier transform
$$\frac{f(0)}{x} + \frac{1}{x} \int_0^\infty f'(t) \cos xt \, dt.$$

The cancelation property implies $f(0) = 0$. Applying (1.61) to the cosine Fourier transform of f', we get the first part of the theorem.

For the cosine Fourier transform, integration by parts implies
$$\widehat{f_c}(x) = -\frac{1}{x} \int_0^\infty f'(t) \sin xt \, dt. \tag{2.23}$$

Since we assume f' to be integrable, its sine Fourier transform is well-defined. We then apply the inverse formula (see, e.g., [100, Vol.1, (4.18) or (5.46)]):
$$-\frac{1}{x} \int_0^\infty f'(t) \sin xt \, dt = \frac{1}{x} \int_0^\infty \mathcal{H}_e(\mathcal{H}_o f')(t) \sin xt \, dt. \tag{2.24}$$

Certain remarks are in order as in the previous section. One must be sure that both Hilbert transforms are well-defined. There is no problem with $\mathcal{H}_o f'$ since

2.3. Derivative in $H_e^1(\mathbb{R}_+)$

f' is integrable. Then we can proceed as in the proof of Theorem 2.8. Applying again (2.15), we cannot follow all the lines as in the proof of Theorem 2.8, since $\frac{1}{u}$ cancels there in the odd Hilbert transform, while for the even one this does not happen. However, there is an even simpler procedure, in fact, applicable in the previous case as well. Denoting again the Hardy type operator of g by

$$\Phi(u) = \frac{1}{u}\int_0^u g(s)\,ds$$

and considering

$$\mathcal{H}_e\Phi(t) = \frac{2t}{\pi}\int_0^\infty \frac{1}{t^2-u^2}\Phi(u)\,du$$

$$= \frac{2t}{\pi}\int_0^\infty \frac{1}{t^2-u^2}\int_0^1 g(su)\,ds\,du,$$

we can see that

$$\mathcal{H}_e\Phi(t) = \int_0^1 \mathcal{H}_e g(st)\,ds = \frac{1}{t}\int_0^t \mathcal{H}_e g(s)\,ds.$$

This means that the Hilbert transform (the even transform here, but it can easily be checked for the general one) of the Hardy operator is the Hardy operator of the Hilbert transform. Since we consider this for integrable functions g ($g = f'$ in our case) with integrable $\mathcal{H}_e g$, both the Hilbert transform and the Hardy operator are well-defined.

Applying on the right-hand side of (2.24) the Hilbert transform $\mathcal{H}_e$ to $\sin xt$ rather than to $\mathcal{H}_o f'$, we have

$$\frac{1}{x}\int_0^\infty \frac{2}{\pi}\int_0^\infty \frac{t(\mathcal{H}_o f')(u)}{t^2-u^2}\,du\,\sin xt\,dt$$

$$= -\frac{1}{x}\int_0^\infty (\mathcal{H}_o f')(u)\left[\frac{2}{\pi}\int_0^\infty \frac{t\sin xt}{u^2-t^2}\,dt\right]du$$

$$= -\frac{1}{x}\int_0^\infty (\mathcal{H}_o f')(u)(\mathcal{H}_o \sin(x\cdot))(u)\,du.$$

The Hilbert transform of $\sin xt$ is $-\cos xt$ (see Example 1.25 or, e.g., [100, Vol.2, Table 1.5 (5.2)]), which turns (2.23) into

$$\frac{1}{x}\int_0^\infty (\mathcal{H}_o f')(t)\cos xt\,dt. \tag{2.25}$$

There is an obstacle in the immediate application of (1.61), since we have $\mathcal{H}_o$ in the last integral, while $\mathcal{H}_e$ is needed. Thus, (1.60) comes into play in (2.25), with f' in place of g. It delivers the needed Hilbert transform $\mathcal{H}_e$ into the Fourier transform in (2.25) and leaves

$$-\frac{1}{x}\int_0^\infty \frac{2}{\pi}\int_0^\infty \frac{f'(u)}{u+t}\,du\cos xt\,dt = -\frac{2}{\pi x}\int_0^\infty f'(u)\int_0^\infty \frac{\cos xt}{u+t}\,dt\,du$$

to be estimated.

First, observing that integration by parts yields

$$\int_0^\infty \frac{\cos xt}{t+u}\,dt = O\left(\frac{1}{xu}\right),$$

we obtain

$$\int_0^\infty \frac{2}{\pi x}\left|\int_{\frac{\pi}{2x}}^\infty f'(u)\int_0^\infty \frac{\cos xt}{u+t}\,dt\,du\right|dx$$

$$\leq \frac{2}{\pi}\int_0^\infty \frac{1}{x^2}\int_{\frac{\pi}{2x}}^\infty \frac{|f'(u)|}{u}\,du\,dx$$

$$= \frac{2}{\pi}\int_0^\infty \frac{|f'(u)|}{u}\,du\int_{\frac{\pi}{2u}}^\infty \frac{1}{x^2}\,dx = \frac{4}{\pi^2}\int_0^\infty |f'(u)|\,du.$$

Hence we proceed to

$$-\frac{2}{\pi x}\int_0^{\frac{\pi}{2x}} f'(u)\int_0^\infty \frac{\cos xt}{u+t}\,dt\,du.$$

Note first that

$$\int_0^\infty \frac{\cos xt}{u+t}\,dt = \cos xu\int_u^\infty \frac{\cos xt}{t}\,dt + \sin xu\int_u^\infty \frac{\sin xt}{t}\,dt.$$

Since the integral in the second term on the right is uniformly bounded, we get

$$\int_0^\infty \frac{2}{\pi x}\left|\int_0^{\frac{\pi}{2x}} f'(u)\sin xu\int_u^\infty \frac{\sin xt}{t}\,dt\,du\right|dx$$

$$\lesssim \int_0^\infty \int_0^{\frac{\pi}{2x}} u|f'(u)|\,du\,dx \lesssim \int_0^\infty |f'(u)|\,du. \qquad (2.26)$$

What remains is

$$-\frac{2}{\pi x}\int_0^{\frac{\pi}{2x}} f'(u)\cos xu\int_{xu}^\infty \frac{\cos t}{t}\,dt\,du.$$

Denoting

$$B = \int_{\frac{\pi}{2}}^\infty \frac{\cos t}{t}\,dt,$$

2.3. Derivative in $H_e^1(\mathbb{R}_+)$

we consider
$$-\frac{2B}{\pi x}\int_0^{\frac{\pi}{2x}} f'(u)\cos xu\, du.$$

Since
$$-\frac{2B}{\pi x}\int_0^{\frac{\pi}{2x}} f'(u)(\cos xu - 1)\, du$$
is estimated exactly as (2.26), we approach to the leading terms by
$$-\frac{2B}{\pi x}\int_0^{\frac{\pi}{2x}} f'(u)\, du = -\frac{2B}{\pi x} f\left(\frac{\pi}{2x}\right). \tag{2.27}$$

Now, our target is reduced to
$$-\frac{2}{\pi x}\int_0^{\frac{\pi}{2x}} f'(u)\cos xu \int_{xu}^{\frac{\pi}{2}} \frac{\cos t}{t}\, dt\, du.$$

Integrating by parts, we get
$$\int_{xu}^{\frac{\pi}{2}} \frac{\cos t}{t}\, dt = O\left(\frac{1}{xu}\right).$$

This and $\cos xu - 1 = O(x^2 u^2)$ imply
$$-\frac{2}{\pi x}\int_0^{\frac{\pi}{2x}} f'(u)(\cos xu - 1)\int_{xu}^{\frac{\pi}{2}} \frac{\cos t}{t}\, dt\, du = O\left(\int_0^{\frac{\pi}{2x}} u|f'(u)|\, du\right),$$

which ends up exactly as (2.26). It remains to consider
$$-\frac{2}{\pi x}\int_0^{\frac{\pi}{2x}} f'(u)\int_{xu}^{\frac{\pi}{2}} \frac{\cos t}{t}\, dt\, du.$$

We have
$$\int_{xu}^{\frac{\pi}{2}} \frac{\cos t}{t}\, dt = \int_{xu}^{\frac{\pi}{2}} \frac{\cos t - 1}{t}\, dt + \ln\frac{\pi}{2ux}.$$

Since
$$\int_0^{xu} \frac{\cos t - 1}{t}\, dt = O(xu),$$
which again leads to the estimate (2.26). It finally remains to consider
$$-\frac{2}{\pi x}\int_0^{\frac{\pi}{2x}} f'(u)\left(\int_0^{\frac{\pi}{2}} \frac{\cos t - 1}{t}\, dt + \ln\frac{\pi}{2ux}\right) du.$$

Along with (2.27) this completes the proof of the theorem. $\square$

2.4 Derivative in a subspace of $H_o^1(\mathbb{R}_+)$ or $H_e^1(\mathbb{R}_+)$

Versions of Theorem 2.8 for various subspaces of $H_0^1(\mathbb{R}_+)$ convenient in applications have been studied repeatedly. The embeddings (see (1.93))

$$O_\infty \hookrightarrow O_{p_1} \hookrightarrow O_{p_2} \hookrightarrow H_o^1 \hookrightarrow L^1 \quad (p_1 > p_2 > 1)$$

are basically reduced to the M. Riesz theorem. A similar embedding for sequences was essentially proved as early as in [59] with the use of the discrete Hilbert transform (with a reference to Stechkin); in explicit form, the M. Riesz theorem was applied in [107]. The operations were carried out with the transform (1.94). Let us show that in this case one does not have to abandon the classical (odd) Hilbert transform either, even though the tails must be estimated. Indeed, we wish to check that

$$\int_0^\infty \left| \int_0^\infty \frac{tg(t)}{x^2 - t^2} dt \right| dx$$

for an integrable function g, which is equivalent to

$$\int_0^\infty \frac{1}{u} \int_u^{2u} \left| \int_0^\infty \frac{tg(t)}{x^2 - t^2} dt \right| dx\, du,$$

is dominated by the norm of g in any of O_q, $1 < q \leq \infty$. Let us carry out the estimates for the cases in which the integral whose absolute value is taken ranges in $[0, \frac{u}{2}]$ or $[2u, \infty)$. The change of integration order gives the inequalities

$$\int_0^\infty \int_0^{\frac{x}{4}} t|g(t)|\, dt \int_{\frac{x}{2}}^x \frac{du}{u(x^2 - t^2)}\, dx$$

$$+ \int_0^\infty \int_{\frac{x}{4}}^{\frac{x}{2}} t|g(t)|\, dt \int_{2t}^x \frac{du}{u(x^2 - t^2)}\, dx$$

$$\lesssim \int_0^\infty t|g(t)| \int_{4t}^\infty \frac{dx}{x^2}\, dt \lesssim \int_0^\infty |g(t)|\, dt$$

in the first case, and the inequalities

$$\int_0^\infty \int_x^{2x} t|g(t)|\, dt \int_{\frac{x}{2}}^{\frac{t}{2}} \frac{du}{u(t^2 - x^2)}\, dx$$

$$+ \int_0^\infty \int_{2x}^\infty t|g(t)|\, dt \int_{\frac{x}{2}}^x \frac{du}{u(t^2 - x^2)}\, dx$$

$$\lesssim \int_0^\infty \int_x^{2x} |g(t)|\, dt \int_{\frac{x}{2}}^{\frac{t}{2}} \frac{du}{u^2}\, dx \lesssim \int_0^\infty \frac{1}{x} \int_x^{2x} |g(t)|\, dt\, dx$$

in the second case, where the last integral is equivalent to $\int_0^\infty |g(t)|\, dt$. It remains to estimate

$$\int_0^\infty \frac{1}{u} \int_u^2 u \left| \int_{\frac{u}{2}}^{2u} \frac{tg(t)}{x^2 - t^2} dt \right| dx\, du.$$

2.4. Derivative in a subspace

It is here that the theory of Hardy spaces comes into play. Let g_u be the function equal to $g(t)$ on $[\frac{u}{2}, 2u]$ and zero outside this interval. By Hölder's inequality,

$$\int_0^\infty \frac{1}{u} \int_u^2 u \left| \int_{\frac{u}{2}}^{2u} \frac{tg(t)}{x^2 - t^2} dt \right| dx\, du$$

$$= \int_0^\infty \frac{1}{u} \int_u^2 u |\mathcal{H}g_u(x)|\, dx\, du$$

$$\leq \int_0^\infty \frac{1}{u} \left(\int_\mathbb{R} |\mathcal{H}g_u(x)|^q\, dx \right)^{\frac{1}{q}} \left(\int_{\frac{u}{2}}^{2u} du \right)^{1-\frac{1}{q}} du.$$

By the M. Riesz theorem, the right-hand side is equivalent to

$$\int_0^\infty u^{-\frac{1}{q}} \left(\int_\mathbb{R} |g_u(x)|^q\, dx \right)^{\frac{1}{q}} du,$$

which is, in turn, equivalent to $\|g\|_{O_q}$.

By this, we have proved the following theorem.

Theorem 2.28. *Let $f \in BV_0[0, \infty) \cap LAC(0, \infty)$ and $f' \in O_q(\mathbb{R}_+)$ for some $1 < q \leq \infty$. Then the cosine Fourier transform of f is integrable, with*

$$\|\widehat{f}_c\|_{L^1(\mathbb{R}_+)} \lesssim \|f'\|_{O_q(\mathbb{R}_+)};$$

while for the sine Fourier transform an asymptotic formula holds: for $x > 0$,

$$\widehat{f}_s(x) = \frac{1}{x} f\left(\frac{\pi}{2x}\right) + F(x),$$

where

$$\|F\|_{L^1(\mathbb{R}_+)} \lesssim \|f'\|_{O_q(\mathbb{R}_+)}.$$

On the other hand, the Riesz theorem says that in essence there are no independent spaces H^q, $1 < q < \infty$; more precisely, they coincide with the spaces L^q. It is no wonder that the only nontrivial technique involved in the direct proof of the result with O_q is the Hausdorff–Young inequality (see, e.g., [67]), which is not related to the theory of Hardy spaces. For completeness, we will give it here.

Direct proof of Theorem 2.28. For simplicity, let us prove the first part. The estimate of the remainder is the same in each case; we have already seen how the leading term in the second part appears. What we should estimate is

$$\int_0^\infty \frac{1}{x} \left| \int_0^\infty f'(t) \sin xt\, dt \right| dx$$

$$= \int_0^\infty \frac{1}{x} \left| \frac{1}{\ln 2} \int_0^\infty \frac{1}{u} \int_u^2 u f'(t) \sin xt\, dt\, du \right| dx.$$

By Fubini, we deal with

$$\int_0^\infty \int_0^\infty \left| \frac{1}{u} \int_u^2 uf'(t) \sin xt\, dt \right| \frac{dx}{x}\, du.$$

In fact, we have to estimate only the inner integrals

$$\int_0^\infty \left| \frac{1}{u} \int_u^2 uf'(t) \sin xt\, dt \right| \frac{dx}{x}$$

$$= \left(\int_0^{\frac{1}{u}} + \int_{\frac{1}{u}}^\infty \right) \left| \frac{1}{u} \int_u^2 uf'(t) \sin xt\, dt \right| \frac{dx}{x}.$$

The first integral on the right is estimated in an elementary way by

$$2 \int_0^{\frac{1}{u}} \int_u^2 u|f'(t)|\, dt\, dx = \frac{2}{u} \int_u^2 u|f'(t)|\, dt,$$

which is twice

$$\left(\frac{1}{u} \int_u^2 u|f'(t)|^q\, dt \right)^{\frac{1}{q}} \tag{2.29}$$

by Hölder's inequality.

As for the remaining integral, we apply Hölder's inequality immediately, which yields

$$\int_{\frac{1}{u}}^\infty \left| \frac{1}{u} \int_u^2 uf'(t) \sin xt\, dt \right| \frac{dx}{x}$$

$$\leq \frac{1}{u} \left(\int_{\frac{1}{u}}^\infty x^{-q}\, dx \right)^{\frac{1}{q}} \left(\int_0^\infty \left| \int_u^2 uf'(t) \sin xt\, dt \right|^{q'} dx \right)^{\frac{1}{q'}}.$$

Calculating the first integral on the right-hand side and applying the Hausdorff-Young inequality (1.17) to the second one provided $1 < q \leq 2$, we again arrive at (2.29) times some constant depending on q. For $q > 2$, the result follows by embeddings (1.93), including $q = \infty$. □

We are now in a position to easily derive the mentioned Pólya's result.

Corollary 2.30. *Let f be a real-valued bounded convex continuous function on $\mathbb{R}_+$, vanishing at infinity. Then $\widehat{f_c} \in L^1(\mathbb{R}_+)$.*

Proof. By convexity and boundedness, such a function is of bounded variation. We mention that within every (a, b) the function, being convex, is Lipschitz (see, e.g., [151]) and thus locally absolutely continuous. Its derivative is integrable on $\mathbb{R}_+$. Since this derivative is monotone decreasing, it also belongs to O_∞, and Theorem 2.28 applies. □

2.4. Derivative in a subspace

It is worth noting that this theorem gets a probabilistic meaning if one assumes $f(0) = 1$. Then the assertion of the theorem can be reformulated as that f is a characteristic function.

The next result follows from the definition of E_q spaces and Theorem 2.20.

Theorem 2.31. *Let $f \in BV_0[0, \infty) \cap LAC(0, \infty)$ and $f' \in E_q(\mathbb{R}_+)$ for some $1 < q \leq \infty$. Then the sine Fourier transform of f is integrable, with*

$$\|\widehat{f_s}\|_{L^1(\mathbb{R}_+)} \lesssim \|f'\|_{E_q(\mathbb{R}_+)};$$

while for the cosine Fourier transform an asymptotic formula holds: for $x > 0$,

$$\widehat{f_c}(x) = -\frac{2}{\pi x} \int_0^{\frac{\pi}{2x}} \frac{f(u)}{u} \, du + F(x), \qquad (2.32)$$

where

$$\|F\|_{L^1(\mathbb{R}_+)} \lesssim \|f'\|_{E_q(\mathbb{R}_+)}.$$

Proof. The reason why we have only one leading term in (2.32), contrary to Theorem 2.20, is that the integral of the first one coincides with that in the definition of E_q if we take $g(t) = f'(t)$ and recall that $f(0) = 0$. For completely the same reason (cf. Remark 2.22), we can integrate by parts in

$$\int_0^{\frac{\pi}{2x}} f'(u) \ln \frac{2ux}{\pi} \, du,$$

where the integrated terms vanish and the integral becomes the one in (2.32).

Estimates for the remainder terms by means of the M. Riesz theorem go along the same lines as above. □

Remark 2.33. It is worth noting that the definition of the E_q classes is consistent with the preceding results in Theorem 2.28. Indeed, the fact that the derivative belongs to this class ensures the integrability of the sine Fourier transform in the first part of Theorem 2.31 and this is exactly what provides the integrability of the sine Fourier transform in the second part of Theorem 2.28.

In order to apply the M. Riesz theorem directly, we arrive at a somewhat different family of subspaces of $H_e^1(\mathbb{R}_+)$. We define, for $1 < p < \infty$, the space $Q_p := Q_p(\mathbb{R}_+)$ to be the subspace of functions g in O_p such that $tg(t)$ is p-integrable on every $[x, \infty)$, $x > 0$.

Theorem 2.34. *Let $f \in BV_0[0, \infty) \cap LAC(0, \infty)$ and $f' \in Q_p(\mathbb{R}_+)$ for some $1 < p < \infty$. Then the sine Fourier transform of f is integrable, with*

$$\|\widehat{f_s}\|_{L^1(\mathbb{R}_+)} \lesssim \|f'\|_{Q_p(\mathbb{R}_+)};$$

while for the cosine Fourier transform an asymptotic formula holds: for $x > 0$,

$$\widehat{f_c}(x) = \frac{A}{x} f\left(\frac{\pi}{2x}\right) + \frac{2}{\pi x} \int_0^{\frac{\pi}{2x}} f'(u) \ln \frac{2ux}{\pi} \, du + F(x),$$

where
$$A = \frac{2}{\pi}\left(\int_0^{\frac{\pi}{2}} \frac{1-\cos t}{t}\,dt - \int_{\frac{\pi}{2}}^{\infty} \frac{\cos t}{t}\,dt\right)$$

and
$$\|F\|_{L^1(\mathbb{R}_+)} \lesssim \|f'\|_{Q_p(\mathbb{R}_+)}.$$

Proof. To apply Theorem 2.20, we only need to prove that $Q_p(\mathbb{R}_+) \subset H_e^1(\mathbb{R}_+)$. Let $g \in H_e^1(\mathbb{R}_+)$. Since $\int_0^\infty g(t)\,dt = 0$, we have

$$\mathcal{H}_e g(x) = \frac{2}{\pi}\int_0^\infty \left[\frac{xg(t)}{x^2-t^2} - \frac{g(t)}{x}\right] dt = \frac{2}{\pi x}\int_0^\infty \frac{t^2 g(t)}{x^2-t^2}\,dt.$$

We observe that, denoting $G(t) = tg(t)$,

$$\mathcal{H}_e g(x) = \frac{1}{x}\mathcal{H}_o G(x).$$

First, we have
$$\int_0^\infty \frac{1}{x}\left|\int_0^{\frac{x}{2}} \frac{t^2 g(t)}{x^2-t^2}\,dt\right| dx \lesssim \int_0^\infty |g(t)|\,dt.$$

Further,
$$\ln 2 \int_0^\infty \frac{1}{x}\left|\int_{\frac{x}{2}}^\infty \frac{t^2 g(t)}{x^2-t^2}\,dt\right| dx \leq \int_0^\infty \frac{1}{x^2}\int_x^{2x}\left|\int_{\frac{u}{2}}^\infty \frac{t^2 g(t)}{u^2-t^2}\,dt\right| du\,dx.$$

Since
$$\int_0^\infty \frac{1}{x^2}\int_x^{2x}\left|\int_{\frac{u}{2}}^\infty \frac{t^2 g(t)}{u^2-t^2}\,dt\right| du\,dx$$
$$\leq \int_0^\infty \frac{1}{x^2}\int_{\frac{x}{2}}^x t^2|g(t)|\int_{2t}^{2x}\frac{du}{u^2-t^2}\,dt\,dx$$
$$\lesssim \int_0^\infty \frac{1}{x^2}\int_{\frac{x}{2}}^x |g(t)|(2x-2t)\,dt\,dx$$
$$\lesssim \int_0^\infty \frac{1}{x}\int_{\frac{x}{2}}^x |g(t)|\,dt\,dx \lesssim \int_0^\infty |g(t)|\,dt,$$

denoting now
$$G_x(t) = \begin{cases} tg(t), & \frac{x}{2} < t < \infty, \\ 0, & \text{otherwise,} \end{cases}$$

2.4. Derivative in a subspace

we have

$$\int_0^\infty \frac{1}{x^2} \int_x^{2x} \left| \int_{\frac{x}{2}}^\infty \frac{t^2 g(t)}{u^2 - t^2} dt \right| du\, dx = \frac{\pi}{2} \int_0^\infty \frac{1}{x^2} \int_x^{2x} |\mathcal{H}_o G_x(u)| \, du\, dx.$$

Applying Hölder's inequality to the inner integral and then applying the M. Riesz theorem, we obtain the upper bound

$$\int_0^\infty \frac{1}{x^2} \left(\int_0^\infty |\mathcal{H}_o G_x(u)|^p \, du \right)^{\frac{1}{p}} x^{\frac{1}{p'}} dx$$

$$\lesssim \int_0^\infty \frac{1}{x^{1+\frac{1}{p}}} \left(\int_x^\infty |tg(t)|^p \, dt \right)^{\frac{1}{p}} dx.$$

Now,

$$\int_0^\infty \frac{1}{x^{1+\frac{1}{p}}} \left(\int_x^\infty |tg(t)|^p \, dt \right)^{\frac{1}{p}} dx \leq \int_0^\infty \frac{1}{x^{1+\frac{1}{p}}} \left(\int_{\frac{x}{2}}^x |tg(t)|^p \, dt \right)^{\frac{1}{p}} dx$$

$$+ \int_0^\infty \frac{1}{x^{1+\frac{1}{p}}} \left(\int_{\frac{x}{2}}^\infty |tg(t)|^p \, dt \right)^{\frac{1}{p}} dx.$$

Since the last term on the right equals the left-hand side times $2^{-\frac{1}{p}}$, we have

$$\int_0^\infty \frac{1}{x^{1+\frac{1}{p}}} \left(\int_x^\infty |tg(t)|^p \, dt \right)^{\frac{1}{p}} dx$$

$$\lesssim \int_0^\infty \frac{1}{x^{1+\frac{1}{p}}} \left(\int_x^{2x} |tg(t)|^p \, dt \right)^{\frac{1}{p}} dx \leq \|g\|_{O_p}.$$

This completes the proof. □

In conclusion, let us remark that if the condition

$$\int_0^\infty \frac{|f(t)|}{t} dt < \infty$$

does not look nice, one can avoid it by sacrificing the sharpness a little.

Lemma 2.35. *Let f be an odd function on $\mathbb{R}$, locally absolutely continuous on $\mathbb{R}_+ \setminus \{0\}$ and such that $\lim_{t \to \infty} f(t) = 0$. Then for $1 < p \leq \infty$*

$$\int_0^\infty \frac{|f(t)|}{t} dt \leq \frac{3}{\ln 2} \int_0^\infty \left(\frac{1}{t} \int_t^{2t} |f'(s)|^p ds \right)^{\frac{1}{p}} |\ln t| \, dt. \qquad (2.36)$$

Proof. We have

$$\int_0^\infty \frac{|f(t)|}{t}\,dt = \left(\int_0^1 + \int_1^\infty\right)\frac{|f(t)|}{t}\,dt = I_1 + I_2.$$

To estimate I_2, we obtain

$$|f(t)| \leq \int_t^\infty |f'(s)|\,ds$$

$$\leq \frac{1}{\ln 2}\int_{\frac{t}{2}}^\infty \frac{1}{s}\int_s^{2s} |f'(u)|\,du\,ds$$

$$\leq \frac{1}{\ln 2}\int_{\frac{t}{2}}^\infty \left(\frac{1}{s}\int_s^{2s} |f'(u)|^p\,du\right)^{\frac{1}{p}}\,ds,$$

and, by Fubini,

$$I_2 \leq \frac{1}{\ln 2}\int_1^\infty \frac{1}{t}\int_{\frac{t}{2}}^\infty \left(\frac{1}{s}\int_s^{2s} |f'(u)|^p\,du\right)^{\frac{1}{p}}\,ds\,dt$$

$$\leq \frac{2}{\ln 2}\int_{\frac{1}{2}}^\infty \left(\frac{1}{s}\int_s^{2s} |f'(u)|^p\,du\right)^{\frac{1}{p}} |\ln s|\,ds. \qquad (2.37)$$

To estimate I_1, we take into account that $f(0) = 0$. As above,

$$|f(t)| \leq \int_0^t |f'(s)|\,ds$$

$$\leq \frac{1}{\ln 2}\int_0^t \frac{1}{s}\int_s^{2s} |f'(u)|\,du\,ds$$

$$\leq \frac{1}{\ln 2}\int_0^t \left(\frac{1}{s}\int_s^{2s} |f'(u)|^p\,du\right)^{\frac{1}{p}}\,ds,$$

and, by Fubini,

$$I_1 \leq \frac{1}{\ln 2}\int_0^1 \frac{1}{t}\int_0^t \left(\frac{1}{s}\int_s^{2s} |f'(u)|^p\,du\right)^{\frac{1}{p}}\,ds\,dt$$

$$\leq \frac{1}{\ln 2}\int_0^1 \left(\frac{1}{s}\int_s^{2s} |f'(u)|^p\,du\right)^{\frac{1}{p}} |\ln s|\,ds. \qquad (2.38)$$

Combining (2.38) and (2.37), we complete the proof. $\square$

It is worth giving separately the following important case.

2.5. Functions on the whole axis

Corollary 2.39. *Let f be an odd function on $\mathbb{R}$, locally absolutely continuous on $\mathbb{R}_+ \setminus \{0\}$ and such that $\lim_{t\to\infty} f(t) = 0$. Then*

$$\int_0^\infty \frac{|f(t)|}{t}\,dt \le \frac{3}{\ln 2}\int_0^\infty |\ln t|\operatorname{ess\,sup}_{s\ge t}|f'(s)|\,dt. \qquad (2.40)$$

In many cases this logarithmic "price" is not essential.

2.5 Functions on the whole axis

In fact, it is easy to imagine a very simple way to obtain a reasonable sufficient condition for the integrability of the Fourier transform of a function of bounded variation on the whole real axis. One should just integrate by parts and apply the Fourier–Hardy inequality (1.61), which gives the mentioned condition as $f' \in H^1(\mathbb{R})$. Of course, the condition at infinity becomes $\lim_{|t|\to\infty} f(t) = 0$. However, there are more delicate options, for example, by considering its even and odd part separately and using the obtained results. More precisely, representing

$$\int_{-\infty}^\infty f(t)e^{-ixt}\,dt = 2\int_0^\infty f_1(t)\cos xt\,dt - 2i\int_0^\infty f_2(t)\sin xt\,dt$$

$$= 2\int_0^\infty f_1(t)\cos|x|t\,dt - 2i\operatorname{sign} x\int_0^\infty f_2(t)\sin|x|t\,dt,$$

where

$$f_1(t) = \frac{f(t)+f(-t)}{2}$$

is the even part of f and

$$f_2(t) = \frac{f(t)-f(-t)}{2}$$

is the odd one, we are likely within the scope of the above setting. The function is assumed to be locally absolutely continuous on $\mathbb{R}\setminus\{0\}$. However, observing that f_1' is odd and f_2' is even and that $f' \in H^1(\mathbb{R})$ is equivalent to $f_1' \in H_o^1(\mathbb{R}_+)$ and $f_2' \in H_e^1(\mathbb{R}_+)$, we find ourselves in the first parts of Theorems 2.8 and 2.20, respectively, which adds nothing as compared with the direct condition $f' \in H^1(\mathbb{R})$.

In any case, with the above results in hand, we do not feel ourselves helpless. We now know that existence of asymptotic formulas is first of all the question of the space to which f belongs. Previous considerations give a clear indication to which spaces f_1 and f_2 should belong instead of the above "natural" ones in order to get more subtle results. The appropriate conditions are $f_1' \in H_e^1(\mathbb{R}_+)$ and $f_2' \in H_o^1(\mathbb{R}_+)$. This leads us to introducing a new Hardy type space $\mathfrak{H}^1(\mathbb{R})$ consisting of the integrable functions g, with even and odd parts g_1 and g_2 defined as above, which satisfy $g_1 \in H_o^1(\mathbb{R}_+)$ and $g_2 \in H_e^1(\mathbb{R}_+)$. The result that follows from Theorems 2.8 and 2.20, more precisely, their "asymptotic" second parts, reads as follows.

Theorem 2.41. *Let $f \in BV_0(\mathbb{R}) \cap LAC(\mathbb{R} \setminus \{0\})$ and $f' \in \mathfrak{H}^1(\mathbb{R})$. Then for the Fourier transform, an asymptotic formula holds: for $x \neq 0$,*

$$\widehat{f}(x) = \frac{A}{|x|}\left[f\left(\frac{\pi}{2|x|}\right) + f\left(-\frac{\pi}{2|x|}\right)\right]$$
$$- \frac{i\,\mathrm{sign}x}{|x|}\left[f\left(\frac{\pi}{2|x|}\right) - f\left(-\frac{\pi}{2|x|}\right)\right]$$
$$+ \frac{1}{\pi|x|}\int_0^{\frac{\pi}{2|x|}} [f'(u) + f'(-u)] \ln \frac{2u|x|}{\pi}\,du + F(x),$$

where

$$A = \frac{2}{\pi}\left(\int_0^{\frac{\pi}{2}} \frac{1-\cos t}{t}\,dt - \int_{\frac{\pi}{2}}^{\infty} \frac{\cos t}{t}\,dt\right)$$

and

$$\|F\|_{L^1(\mathbb{R})} \lesssim \|f'\|_{\mathfrak{H}^1(\mathbb{R})}.$$

Of course, there is an alternative way to study the Fourier transform of a function of bounded variation defined on the whole real axis. Instead of considering its even and odd parts, one may try to consider separately the part of the function on the right half-axis and on the left one and check their properties in connection with Theorems 2.8 and 2.20, or other tests. However, in the general case, one should apply many conditions to each of this parts.

2.6 Absolute continuity, integrability of the Fourier transform and a Hardy–Littlewood theorem

We are concerned with extensions of the following classical result due to Hardy and Littlewood (see [81] or, e.g., [211, Vol.I, Ch.VII, (8.6)]; also see [91, Ch. 5] where a version of this theorem is proved by means of the theorem of the brothers Riesz; for an interesting different proof, see [190]).

Theorem 2.42. *If a (periodic) function f and its conjugate $\widetilde{f}$ are both of bounded variation, then their Fourier series converge absolutely.*

The first attempt to transfer these circumstances to the real line was undertaken as long ago as in 1935 (see [88]). However, the result obtained there was far from being optimal, since a very restrictive condition of belonging of the considered function to a Lebesgue space (in addition to being of bounded variation) was assumed, only for the Hilbert transform to exist.

Since this problem is closely related to absolute continuity, let us try a different approach to it. The celebrated theorem of F. and M. Riesz states that if the negative Fourier coefficients of a measure are all zeros, then the measure ia absolutely continuous with respect to the Lebesgue measure. Three different proofs of

2.6. Hardy–Littlewood theorem

this theorem are given by Koosis in [105]; the role and importance of this result are thoroughly discussed as well. Two of these proofs are produced by means of methods of complex analysis, one of them due to Helson and Lowdenslager uses methods of functional analysis. Its non-periodic analog (can be found in [163, Th.8.2.7]) asserts that if

$$\int_{-\infty}^{\infty} e^{-ixt} \, d\mu(t) = 0, \quad \text{for all } x < 0,$$

then μ is absolutely continuous. It is proved by transference to the periodic case.

In both cases the statement is intimately related to the Hardy space H^1. In the periodic case, it is done via Poisson's representation for harmonic functions. However, in the non-periodic case there is a variety of different spaces of such type; for a discussion, see, e.g., Helson's book [85]. One is obtained by boundary values on $\mathbb{R}$ from the upper half-plane after mapping from the circle, for instance, it contains constants. The other is a subspace of $L^1(\mathbb{R})$ whose Fourier transform vanishes for negative arguments. However, we are interested in a different one, defined by means of the Hilbert transform.

All these lead to the following question:

Is it true that if $\mathcal{H}d\mu \in L^1(\mathbb{R})$, then the measure is absolutely continuous with respect to the Lebesgue measure?

We note that due to the classical result of Wiener and Wintner in [205] (numerous researchers continued this study, for example, Ivashev–Musatov in [93]), if one constructs the analog of the left-hand side of the Fourier–Hardy inequality

$$\int_{\mathbb{R}} \left| \int_{\mathbb{R}} e^{ixt} d\mu(t) \right| \frac{dx}{|x|},$$

the mentioned results confirm the existence of a singular measure μ (or a singular function generating a measure) for which this integral is finite.

It is known (see, e.g., [70, Lemma 2]) that if $f \in L^1(\mathbb{R})$ is such that $\widehat{f}(t) = 0$ for $t \leq 0$, then $f \in H^1(\mathbb{R})$ but the converse is not true. Therefore, the above generalization of the theorem of the brothers Riesz cannot help us. However, the answer is affirmative. There is a result due to Khrushchev and Vinogradov (see, e.g., [145] and references therein) that says that the (rearrangement of the) Hilbert transform of a singular measure behaves near infinity roughly as $\frac{1}{x}$ and hence cannot be integrable. A one-dimensional result very similar to Theorem 2.42 has recently been obtained in [132]. It asserts that if f is a function of bounded variation that vanishes at infinity: $\lim_{|t| \to \infty} f(t) = 0$, and if its conjugate (see (1.63))

$$\widetilde{\mathcal{H}}f(x) = (\text{P.V.}) \frac{1}{\pi} \int_{\mathbb{R}} f(t) \left\{ \frac{1}{x-t} + \frac{t}{1+t^2} \right\} dt$$

is also of bounded variation, then the Fourier transforms of both functions are integrable on $\mathbb{R}$. We present a somewhat more general version of this result.

Denoting $IL^1(\mathbb{R})$ to be the class of such functions that each is differentiable almost everywhere and the derivative is integrable over $\mathbb{R}$, we start with the commutativity relations.

Theorem 2.43. Let $f \in L^\infty(\mathbb{R})$ and $\widetilde{\mathcal{H}}f \in IL^1(\mathbb{R})$.

$$\text{If } f \in AC(\mathbb{R}), \text{ we have } \frac{d}{dx}\widetilde{\mathcal{H}}f(x) = \mathcal{H}f'(x) \text{ a.e.} \tag{2.44}$$

$$\text{If } f \in BV(\mathbb{R}), \text{ we have } \frac{d}{dx}\widetilde{\mathcal{H}}f(x) = \mathcal{H}df(x) \text{ a.e..} \tag{2.45}$$

Proof. We do calculations somewhat similar to those in [156, Th. 1]. The main difference is that in that paper direct differentiation is applied to $\mathcal{H}f$, with $f \in C_0^\infty$, while in our case the class of functions involved needs more delicate arguments. We first split the integral

$$\frac{d}{dx}\left[(\text{P.V.})\int_\mathbb{R} f(t)\left\{\frac{1}{x-t} + \frac{t}{1+t^2}\right\}dt\right]$$

$$= \frac{d}{dx}\left[(\text{P.V.})\int_{|x-t|\leq 1} + \int_{|x-t|>1}\right]f(t)\left\{\frac{1}{x-t} + \frac{t}{1+t^2}\right\}dt. \tag{2.46}$$

For the second one, we get

$$\frac{d}{dx}\left[\int_{-\infty}^{x-1} + \int_{x+1}^{\infty}\right]f(t)\left\{\frac{1}{x-t} + \frac{t}{1+t^2}\right\}dt$$

$$= f(x-1)\left\{1 + \frac{x-1}{1+(x-1)^2}\right\} - f(x+1)\left\{-1 + \frac{x+1}{1+(x+1)^2}\right\}$$

$$+ \left[\int_{-\infty}^{x-1} + \int_{x+1}^{\infty}\right]\frac{f(t)}{(x-t)^2}dt.$$

Integrating by parts on the right, we obtain

$$f(x-1)\frac{x-1}{1+(x-1)^2} - f(x+1)\frac{x+1}{1+(x+1)^2} + \int_{|x-t|>1}\frac{f'(t)}{x-t}dt. \tag{2.47}$$

For the first integral on the right-hand side of (2.46), we have

$$\frac{d}{dx}(\text{P.V.})\int_{x-1}^{x+1} f(t)\left\{\frac{1}{x-t} + \frac{t}{1+t^2}\right\}dt$$

$$- f(x-1)\frac{x-1}{1+(x-1)^2} + f(x+1)\frac{x+1}{1+(x+1)^2} - \frac{d}{dx}\int_{|t|<1} F(x,t)\,dt,$$

where $F(x,t) = \frac{f(x+t)-f(x)}{t}$ for $t \neq 0$ (it is possible to replace $f(x+t)$ by $f(x+t) - f(x)$ due to understanding the integral in the principal value sense),

2.6. Hardy–Littlewood theorem

and $F(x, 0) = f'(x)$. The latter takes place a.e. Combining this with (2.47), we rewrite the right-hand side of (2.46) as

$$\int_{|x-t|>1} \frac{f'(t)}{x-t} dt - \frac{d}{dx} \int_{|t|<1} F(x, t) \, dt. \tag{2.48}$$

We deal with the last term on the right-hand side of (2.48) as follows. By Fubini, the integral is

$$\int_{|t|\leq 1} \int_x^{x+t} f'(s) ds \frac{dt}{t} = \int_{x-1}^{x+1} f'(s) \ln \frac{1}{|x-s|} ds. \tag{2.49}$$

By assumption, its derivative exists almost everywhere. If we let the derivative act in the integral, the result also exists almost everywhere as the truncated Hilbert transform of an integrable function. If these two values do not coincide on a set A of positive measure, we apply Egorov's theorem to let the integral on the right-hand side of (2.49) converge in the principal value sense uniformly with respect to x on a set slightly smaller than A. This ensures the commutativity of the differentiation and principal value integration (see, e.g., [100, §4.8]), which contradicts our assumption that they do not coincide on any part of A. This contradiction along with taking into account the not considered yet integral over $|t| > 1$ proves the assertion.

The proof for $\widetilde{\mathcal{H}}f \in BV$ goes along the same lines, just $df(t)$ appears in (2.48) and above instead of $f'(t)$, and $df(s)$ is used in place of $f'(s)$. This completes the proof. □

Remark 2.50. The same results, but with $\mathcal{H}f$ instead of $\widetilde{\mathcal{H}}f$, are true if one assumes f to be Lebesgue integrable.

If a periodic function and its conjugate are both of bounded variation, then the function is absolutely continuous is a well-known result of F. and M. Riesz; see [211, Ch.VII, (8.2)] for a proof by means of complex analysis and [13, Ch. VIII, §12] for a nice discussion. Using this result, we have proved in [132] that if both f and $\widetilde{\mathcal{H}}f$ are of bounded variation, then f is locally absolutely continuous. We give a proof of a somewhat more general assertion.

Theorem 2.51. *Let $f \in BV(\mathbb{R})$. If $\widetilde{\mathcal{H}}f \in IL^1(\mathbb{R})$, then f is absolutely continuous. The same is true for $\widetilde{\mathcal{H}}f$ provided it is only of bounded variation.*

Proof. By the previous proposition, $\frac{d}{dx}\widetilde{\mathcal{H}}f(x) = \mathcal{H}df(x)$ a.e. By assumption,

$$\mathcal{H}df(x) \in L^1(\mathbb{R}).$$

As is mentioned above, such a function must be absolutely continuous; otherwise the function $x \to |\{|\mathcal{H}df > x\}|$ behaves, roughly speaking, as $\frac{1}{|x|}$ near infinity. In fact, in a simpler version it goes back to an old result due to Boole on the Hilbert transform of a finite linear combination of point masses. See, e.g., [42, Chapter 7], [145] and [161].

Let now $\widetilde{\mathcal{H}}f \in BV$. In virtue of [102, (3.2)],

$$\widetilde{\mathcal{H}}(\widetilde{\mathcal{H}}f)(x) = -f(x) + \frac{1}{\pi}\int_{\mathbb{R}} \frac{f(u)}{1+u^2}\,du; \qquad (2.52)$$

therefore f is, up to a constant and sign, the conjugate of $\widetilde{\mathcal{H}}f$. Putting $g = \widetilde{\mathcal{H}}f$, we have that $g \in BV$ and $\widetilde{\mathcal{H}}g \in IL^1(\mathbb{R})$, perfectly satisfying the previous assumptions of the theorem. Hence the above gives that $\widetilde{\mathcal{H}}f$ is absolutely continuous, as required. □

We are now in a position to present a proof of a somewhat extended version of the Hardy–Littlewood theorem based on the above refined results.

Theorem 2.53. *Let $f \in BV_0$ and $\widetilde{\mathcal{H}}f \in IL^1(\mathbb{R})$. Then the Fourier transform of f is integrable on $\mathbb{R}$. If $\widetilde{\mathcal{H}}f \in BV_0$, then the Fourier transform of $\widetilde{\mathcal{H}}f$ is also integrable on $\mathbb{R}$.*

Proof. Since the function f is of bounded variation, its derivative f' exists almost everywhere and is integrable. It follows from the assumption on $\widetilde{\mathcal{H}}f$ and from Theorem 2.43 that $\mathcal{H}f'(x)$ exists at almost every x and is also integrable. Therefore $f' \in H^1(\mathbb{R})$. We shall now make use of (1.61) with $g = f'$. Observe that the assumptions of the theorem imply the cancelation property (1.49) for f'. Integrating by parts, which is possible since f is absolutely continuous, we obtain

$$\widehat{f'}(x) = \int_{\mathbb{R}} f'(t)e^{-itx}\,dt = ix\int_{\mathbb{R}} f(t)e^{-itx}\,dt,$$

where the integral on the right-hand side is understood in the improper sense. Hence, the left-hand side of (1.61) is exactly the L^1 norm of the Fourier transform of f, which is finite since the right-hand side of (1.61) in this case is the norm of $f' \in H^1(\mathbb{R})$.

Further, we have $i\,\text{sign}\,x\widehat{f'}(x) = \widehat{\mathcal{H}f'}(x)$, which, by Theorem 2.51 and (2.44) in Theorem 2.43, is the Fourier transform of $\dfrac{d}{dx}\widetilde{\mathcal{H}}f$. Integrating by parts as above, we have

$$i\,\text{sign}\,x\widehat{f'}(x) = \lim_{A\to\infty}\int_{-A}^{A} \frac{d}{dt}\widetilde{\mathcal{H}}f(t)e^{-ixt}\,dt$$

$$= \lim_{A\to\infty}\left[\widetilde{\mathcal{H}}f(t)e^{-ixt}\right]_{-A}^{A} + ix\int_{\mathbb{R}}\widetilde{\mathcal{H}}f(t)e^{-ixt}\,dt = ix\widehat{\widetilde{\mathcal{H}}f}(x).$$

Dividing as above by ix gives in the left-hand side an integrable function, whose L^1 norm is dominated by the H^1 norm of f'. Therefore, the L^1 norm of $\widetilde{\mathcal{H}}f$ is also finite, which completes the proof. □

2.6. Hardy–Littlewood theorem

This means that if we wish to deal with wide subspaces of the class BV, like in the previous sections of this chapter, we must stay in the world of locally absolutely continuous functions, as before. Therefore, we will continue to deal with such functions and will not return to this question in the multidimensional setting. However, the other results of this section will be generalized.

Observe also that in virtue of Remark 1.10 this theorem proves, in fact, the assertion in full generality, that is, for general finite Borel measures.

Chapter 3
Integrability spaces: wide, wider and widest

In this chapter, we discuss how wide a space for the integrability of the Fourier transform of a function of bounded variation can be. Of course, we continue to make use of the natural assumption of local absolute continuity. Also, our functions vanish at infinity. What is naturally related to this is, on the one hand, a very general asymptotic formula for the sine Fourier transform, and, on the other hand, a variety of subspaces of the class of functions of bounded variation each of them has its own "integrability position". It will be shown that all these are connected to various versions and refinements of the Fourier-Hardy inequality (1.61). This chapter mainly refers to the recent publications [118] and [121]. It is worth noting that an asymptotic formula for the sine Fourier transform of **all** $f \in BV_0[0, \infty) \cap LAC(0, \infty)$ in Subsection 4.2 is one of the central results of the whole book. In all the other asymptotic formulas throughout the book, the function (or, more precisely, its derivative) is subject to certain additional restrictions.

3.1 Widest integrability spaces

As we have seen, the spaces $H_o^1(\mathbb{R}_+)$, $H_e^1(\mathbb{R}_+)$ and $A_{1,2}$ look sufficiently large to be close enough to the widest possible spaces to which the belonging of the derivative f' ensures the integrability of the cosine and sine Fourier transforms of f. However, the possibility of existence of a wider space of such type is of considerable interest; moreover, it is interesting whether the widest space of such type can be pointed out.

Clarifying it at once, let us show that there is a way to indicate the maximal space of integrability. In fact, it has in essence been introduced (for different purposes) by Johnson and Warner in [96] as

Chapter 3. Integrability spaces: wide, wider and widest

Definition 3.1.
$$Q = \{g : g \in L^1(\mathbb{R}), \int_{\mathbb{R}} \frac{|\widehat{g}(x)|}{|x|} dx < \infty\}.$$

With the obvious norm (cf. (1.85))

$$\|g\|_{L^1(\mathbb{R})} + \int_{\mathbb{R}} \frac{|\widehat{g}(x)|}{|x|} dx,$$

it is a Banach space and ideal in $L^1(\mathbb{R})$.

The following result is a simple effect of the structure of this space (see [118]).

Theorem 3.2. *Let* $f : \mathbb{R}_+ \to \mathbb{C}$ *be locally absolutely continuous on* $(0, \infty)$, *of bounded variation and* $\lim_{t \to \infty} f(t) = 0$.

a) *The cosine Fourier transform of* f *given by (2.21) is Lebesgue integrable on* $\mathbb{R}_+$ *if and only if* $f' \in Q$.

b) *The sine Fourier transform of* f *given by (2.10) is Lebesgue integrable on* $\mathbb{R}_+$ *if and only if* $f' \in Q$.

Discussion and comments are in order. First, Theorem 3.2 does not seem to be a result at all, rather at most a technical reformulation of the definitions (1.13) and (1.14). This could have been so but not after the appearance of the analysis of Q in [96]. Indeed, (1.61) implies

$$H^1(\mathbb{R}) \subseteq Q \subseteq L^1_0(\mathbb{R}). \tag{3.3}$$

Recall that the latter is the subspace of all the functions g in $L^1(\mathbb{R})$ which satisfy the cancelation property (1.49). Such functions are sometimes called wavelet functions, see, e.g., [164].

In fact, in a) and b) of Theorem 3.2 the conditions are given in terms of different subspaces of Q. We shall use the space Q_o of the odd functions from Q:

$$Q_o = \left\{g : g \in L^1(\mathbb{R}), g(-t) = -g(t), \int_0^\infty \frac{|\widehat{g}_s(x)|}{x} dx < \infty\right\};$$

such functions naturally satisfy (1.49). This is exactly the space that works in a). As for b), an even counterpart of Q_o is used, more precisely,

$$Q_e = \left\{g : g \in L^1(\mathbb{R}), g(-t) = g(t), \int_0^\infty \frac{|\widehat{g}_c(x)|}{x} dx < \infty\right\}.$$

This makes sense only if (1.80) holds.

Theorem 3.2 yields a simple consequence for the Fourier transform on the whole axis (cf. [68]).

3.2. The sine Fourier transform

Corollary 3.4. *Let $f : \mathbb{R} \to \mathbb{C}$ be locally absolutely continuous on $\mathbb{R}\setminus\{0\}$, of bounded variation and $\lim_{|t|\to\infty} f(t) = 0$. Then $\widehat{f}$ is Lebesgue integrable on $\mathbb{R}$ if and only if $f(0) = 0$ and $f' \in Q$.*

On the one hand, the obtained conditions are necessary and sufficient, which is a weighty argument (for the latter, I like the German word Totschlagargument). On the other hand, it does not look convenient to be applied as a test. Therefore, in the next sections we proceed to subspaces of this space, first of all, wider than $H_o^1(\mathbb{R}_+)$ and $H_e^1(\mathbb{R}_+)$, and study various interrelations between all these spaces.

3.2 The sine Fourier transform

For the cosine Fourier transform, in this context we cannot suggest more than a) in Theorem 3.2. Of course, the approach in the previous chapter can be used, but this is a different story. The situation is more delicate with the sine Fourier transform, where a sort of asymptotic relation can be obtained for **all** $f \in BV_0[0, \infty) \cap LAC(0, \infty)$. On the one hand, (2.10) and (2.11) do not follow from Theorem 3.2. On the other hand, Theorem 2.8 is already a hint of non-stoppage after getting b) in Theorem 3.2 in order to search for something more developed. Somewhat surprisingly, in a situation more general than that in Theorem 2.8, the form of the answer is different from that in (2.10).

Recalling (1.122), we have

$$\frac{\widehat{g}_s(x)}{x} = B_s g(x).$$

In the following theorem, no assumptions except natural and mild

$$f \in BV_0[0, \infty) \cap LAC(0, \infty)$$

are posed; that is, the result is valid for "all unartificial" functions of bounded variation.

Theorem 3.5. *Let $f : \mathbb{R}_+ \to \mathbb{C}$ be locally absolutely continuous on $(0, \infty)$, of bounded variation and $\lim_{t\to\infty} f(t) = 0$. Then for the sine Fourier transform of f, there holds for any $x > 0$*

$$\widehat{f}_s(x) = \frac{1}{x} f\left(\frac{\pi}{2x}\right) - \mathcal{H}_o B_s f'(x) + G(x), \tag{3.6}$$

where

$$\|G\|_{L^1(\mathbb{R}_+)} \lesssim \|f'\|_{L^1(\mathbb{R}_+)}. \tag{3.7}$$

It is worth discussing the "mysterious" term $\mathcal{H}_o B_s f'(x)$ in the asymptotic formula (3.6). First of all, it reflects the fact that (3.6) is obtained for an arbitrary function of bounded variation (again, local absolute continuity and vanishing at infinity in these circumstances cannot be considered as hard restrictions). This means that the hope that it is separated "by mistake" and can be proved to be integrable is groundless.

To compare with Theorem 3.2, it should be mentioned that the necessary and sufficient condition in a) is, in fact, $f' \in Q_o$. Theorem 3.5 also deals with Q_o but in a more sophisticated way and makes it natural to introduce a special Hardy type space $H_Q^1(\mathbb{R}_+)$.

Definition 3.8. *The space $H_Q^1(\mathbb{R}_+)$ consists of Q_o functions g with integrable $\mathcal{H}_o B_s g$.*

Corollary 3.9. *If a function f satisfies the assumptions of Theorem 3.5 and is such that $f' \in H_Q^1(\mathbb{R}_+)$, then*

$$\widehat{f}_s(x) = \frac{1}{x} f\left(\frac{\pi}{2x}\right) + G(x),$$

where

$$\|G\|_{L^1(\mathbb{R}_+)} \lesssim \|f'\|_{H_Q^1(\mathbb{R}_+)}.$$

Technically, this is an obvious corollary of Theorem 3.5. We shall discuss it later on.

Proof of Theorem 3.5. Let us start with integration by parts in

$$\int_0^{\frac{\pi}{2x}} f(t) \sin xt \, dt = \left.\frac{1-\cos xt}{x} f(t)\right|_0^{\frac{\pi}{2x}} + \frac{1}{x} \int_0^{\frac{\pi}{2x}} f'(t)[\cos xt - 1] \, dt$$

$$= \frac{1}{x} f\left(\frac{\pi}{2x}\right) + \frac{1}{x} \int_0^{\frac{\pi}{2x}} f'(t)[\cos xt - 1] \, dt.$$

The last value is bounded by $\int_0^{\frac{\pi}{2x}} t|f'(t)|\, dt$, and

$$\int_0^\infty \int_0^{\frac{\pi}{2x}} t|f'(t)|\, dt\, dx = \frac{\pi}{2} \int_0^\infty |f'(t)|\, dt. \tag{3.10}$$

Going over to

$$I = I(x) = \int_{\frac{\pi}{2x}}^\infty f(t) \sin xt\, dt,$$

we continue with the following statement.

Lemma 3.11. *If f satisfies the assumptions of Theorem 3.5, then*

$$-\mathcal{H}_o B_s f'(x) = \frac{2}{x\pi} \int_0^\infty f'(t) \sin xt \int_{xt}^\infty \frac{\cos v}{v}\, dv\, dt$$

$$+ \frac{2}{x\pi} \int_0^\infty f'(t) \cos xt \int_0^{xt} \frac{\sin v}{v}\, dv\, dt. \tag{3.12}$$

3.2. The sine Fourier transform

Proof of Lemma 3.11. Using standard notations

$$\text{Ci}(u) = -\int_u^\infty \frac{\cos t}{t}\, dt$$

and

$$\text{Si}(u) = \int_0^u \frac{\sin t}{t}\, dt = \frac{\pi}{2} - \int_u^\infty \frac{\sin t}{t}\, dt,$$

we will make use of the formula (see [15, Ch.II, §2.2 (18)])

$$\int_0^\infty \frac{1}{x^2 - u^2} \sin ut\, du = \lim_{\delta \to 0+} \int_{|x-u|>\delta} \frac{1}{x^2 - u^2} \sin ut\, du$$

$$= \frac{1}{x}[\sin xt\, \text{Ci}(xt) - \cos xt\, \text{Si}(xt)], \qquad (3.13)$$

where the integrals are understood in the principal value sense and $x, t > 0$. Since $f' \in L^1(\mathbb{R}_+)$ and the limit in (3.13) is uniform in t, we have

$$\mathcal{H}_o B_s f'(x) = \frac{2}{\pi} \int_0^\infty \widehat{f_s'}(u) \frac{1}{x^2 - u^2}\, du$$

$$= \frac{2}{\pi} \int_0^\infty f'(t) \int_0^\infty \frac{1}{x^2 - u^2} \sin ut\, du\, dt. \qquad (3.14)$$

Applying (3.13) to the inner integral on the right-hand side of (3.14) and using the expressions for Ci and Si complete the proof of the lemma. □

With this relation in hand, we are going to prove that

$$I(x) = -\mathcal{H}_o B_s f'(x) + G(x), \qquad (3.15)$$

where

$$\|G\|_{L^1(\mathbb{R}_+)} \lesssim \|f'\|_{L^1(\mathbb{R}_+)}.$$

We denote the two summands on the right-hand side of (3.12) by I_1 and I_2. For both, we make use of the fact that

$$\int_{xt}^\infty \frac{\cos v}{v}\, dv = O\left(\frac{1}{xt}\right).$$

The same is true when $\cos v$ is replaced by $\sin v$. We begin with I_1. For $t \geq \frac{1}{x}$, we have

$$\int_0^\infty \frac{1}{x} \int_{\frac{1}{x}}^\infty |f'(t)| \frac{1}{xt}\, dt\, dx = \int_0^\infty \frac{|f'(t)|}{t} \int_{\frac{1}{t}}^\infty \frac{1}{x^2}\, dx\, dt$$

$$= \int_0^\infty |f'(t)|\, dt. \qquad (3.16)$$

For $t \leq \frac{1}{x}$, we split the inner integral in I_1 into two. First,

$$\int_1^\infty \frac{\cos v}{v}\, dv = O(1),$$

and using $\left|\frac{\sin xt}{x}\right| \leq t$, we arrive at the relation similar to (3.10). Further, we have

$$\int_{xt}^1 \left|\frac{\cos v}{v}\right| dv = O\left(\ln \frac{1}{xt}\right).$$

By this, integrating in x over $(0, \infty)$, we end up with

$$\int_0^\infty |f'(t)| t \int_0^{1/t} \ln \frac{1}{xt}\, dx\, dt = \int_0^\infty |f'(t)|\, dt.$$

Here we use that

$$\int_0^{\frac{1}{t}} \ln \frac{1}{xt}\, dx = \frac{1}{t}.$$

In conclusion, I_1 can be treated as G.

Let us proceed to I_2. Using that

$$\frac{1}{x}\left|\int_0^{xt} \frac{\sin v}{v}\, dv\right| = O(t),$$

we arrive, for $t \leq \frac{\pi}{2x}$, at (3.10). Let now $t \geq \frac{\pi}{2x}$. We have

$$\int_0^{xt} \frac{\sin v}{v}\, dv = \frac{\pi}{2} - \int_{xt}^\infty \frac{\sin v}{v}\, dv.$$

Now

$$\frac{2}{x\pi} \int_{\frac{\pi}{2x}}^\infty f'(t) \cos xt\, dt \frac{\pi}{2} = I.$$

For the integral $\int_{xt}^\infty \frac{\sin v}{v}\, dv$, the estimates are exactly like those in (3.16).

Combining (3.15) and the estimates before Lemma 3.11, we complete the proof of the theorem. □

Remark 3.17. Formally, there exists a certain symmetry between the two Fourier transforms. Indeed, using the formula (see [15, Ch.II, §1.2 (15)])

$$\int_0^\infty \frac{1}{a^2 - x^2} \cos yx\, dx = \frac{\pi}{2a} \sin ay,$$

and recalling that

$$B_c f'(x) = \left(\frac{\widehat{f_c'}(u)}{u}\right)(x)$$

(see Section 1.5 in Chapter 1), we obtain

$$\mathcal{H}_o B_c f'(x) = \frac{2}{\pi} \int_0^\infty \frac{u}{x^2 - u^2} \frac{1}{u} \int_0^\infty f'(t) \cos ut \, dt \, du$$

$$= \frac{2}{\pi} \int_0^\infty f'(t) \int_0^\infty \frac{\cos ut}{x^2 - u^2} \, du \, dt$$

$$= \frac{1}{x} \int_0^\infty f'(t) \sin xt \, dt. \qquad (3.18)$$

Comparing this with the proof of a) in Theorem 3.2, on the one hand, and with (3.6) on the other hand, we see that (3.18) represents the cosine Fourier transform in the form of a counterpart of (3.6) in the sense that it gives an analog of the second term on the right-hand side of (3.6), while the "remainder term" is identically zero. Also, this setting does not allow one to reproduce the first leading term, being an ultimate feature of the sine transform in it. The setting where this is possible to a certain extent has been demonstrated in the previous chapter.

3.3 Intermediate spaces

It is highly improbable that Q (or Q_o) can be defined in terms of g itself rather than its Fourier transform. Hence it is of interest to find certain proper subspaces of Q_o, wider than H_o^1, belonging to which can be verified in a reasonable way. For example, in the paper [177] a family of nested subspaces between H^1 and L^1 is introduced and duality properties of that family are studied. However, it is not clear how to compare that family with Q_o. Before searching for new subspaces, the relations between those that already have been considered should be made more exact.

3.3.1 Embeddings

Back to Theorem 3.5, let us analyze (3.15). On the one hand, we have

$$\int_0^\infty |I(x)| \, dx = \int_0^\infty |\mathcal{H}_o B_s f'(x)| \, dx + O(\|f'\|_{H_0^1(\mathbb{R}_+)}).$$

On the other hand, it is proved in [107] that

$$\int_0^\infty |I(x)| \, dx = O(\|f'\|_{H_0^1(\mathbb{R}_+)}).$$

This leads to

Proposition 3.19. *If g is an integrable odd function, then*

$$\|\mathcal{H}_o B_s g\|_{L^1(\mathbb{R}_+)} \lesssim \|g\|_{H_0^1(\mathbb{R}_+)}.$$

The above proof of Proposition 3.19 looks "artificial". Let us give a direct proof.

Proof of Proposition 3.19. The proof will be molecular. We represent g as a molecular sum. To prove the proposition, it suffices to show that, for every molecule M, the function $B_s M$ is also a molecule. Indeed, since we consider the odd Hilbert transform, (1.49) is satisfied automatically for the odd extension of $B_s M$. Further, we have

$$|\widehat{M_s}(x)| \leq \int_0^x |M(t)|\, dt + x \int_x^\infty |M(t)|\, t\, dt$$

and

$$\left(\int_0^\infty \frac{|\widehat{M_s}(x)|^2}{x^2}\, dx\right)^{\frac{1}{2}} \leq \left(\int_0^\infty x^{-2} \left(\int_0^x |M(t)|\, dt\right)^2 dx\right)^{\frac{1}{2}}$$
$$+ \left(\int_0^\infty \left(\int_x^\infty |M(t)|\, t\, dt\right)^2 dx\right)^{\frac{1}{2}} = I_1 + I_2.$$

We will estimate both I_1 and I_2 by means of Hardy's inequalities (see (1.62) or, e.g., [79, (330)]):

$$\left[\int_0^\infty |x^{2b} R(x)|^2\, dx\right]^{\frac{1}{2}} \lesssim \left[\int_0^\infty x^{2b+2} \psi(x)^2\, dx\right]^{\frac{1}{2}},$$

where, for $\psi(x) \geq 0$, either

$$R(x) = \int_0^x \psi(t)\, dt$$

and $2b < -1$, or

$$R(x) = \int_x^\infty \psi(t)\, dt$$

and $2b > -1$. For I_1, we have the first case with $\psi(t) := |M(t)|$ and $2b = -2 < -1$, correspondingly $I_1 \lesssim \|M\|_{L^2(\mathbb{R}_+)}$. For I_2, we have the second case with $\psi(t) := t|M(t)|$, $2b = 0 > -1$, and

$$I_2 \lesssim \left[\int_0^\infty x^2 |M(x)|^2\, dx\right]^{\frac{1}{2}}.$$

Square integrability of $\widehat{M_s}$ immediately follows from Parseval's identity and the square integrability of M, which completes the proof. $\square$

This implies a broadened chain of embeddings:

$$H_0^1(\mathbb{R}_+) \subseteq H_Q^1(\mathbb{R}_+) \subseteq Q_o \subseteq L_0^1(\mathbb{R}_+).$$

It is very interesting to figure out which of these embeddings are proper. Correspondingly, intermediate spaces are of interest, both theoretically and practically.

3.3.2 A counterexample

We will prove that $H_o^1(\mathbb{R}_+) \neq Q_o$. The above considerations reduce this problem to constructing a function of bounded variation such that its cosine Fourier transform is integrable, while its derivative is not in $H_o^1(\mathbb{R}_+)$. Related counterexamples for sequences can be found in [10] and [61]. We will build an example for functions in a similar manner, more precisely, an example of a function g from $A_{1,2}$ (see (1.117)) but not from $H_o^1(\mathbb{R}_+)$ will be constructed (cf. Theorem 3.22 below). First, we may think that g is good enough on $(0,1)$ and concentrate on its behavior on $(1,\infty)$. Therefore, we compare the values

$$N_1 = \int_1^\infty \left| \int_0^{\frac{t}{2}} \frac{g(t-s) - g(t+s)}{s} ds \right| dt$$

and

$$N_2 = \sum_{m=0}^\infty \left\{ \sum_{j=1}^\infty \left[\int_{j2^m}^{(j+1)2^m} |g(t)| \, dt \right]^2 \right\}^{\frac{1}{2}} dx < \infty.$$

Recall that we can restrict ourselves to the upper limit $\frac{t}{2}$ in the inner integral for N_1, since that integral, the T-transform, differs from the Hilbert transform of the odd function g by an integrable function (see (1.95)). Let the function g be non-negative, vanishing everywhere except for the intervals $(2^m, 2^m+1)$ and satisfying

$$\int_{2^m}^{2^m+1} g(t) \, dt = \frac{1}{(m+1)^2}$$

for $m = 0, 1, 2, \ldots$ Obviously, the integral does not vanish only when j is a power of 2, hence

$$N_2 \leq \sum_{m=0}^\infty \left\{ \sum_{j=0}^\infty \frac{1}{(m+1+j)^4} \right\}^{\frac{1}{2}} \leq \sum_{m=0}^\infty (m+1)^{-\frac{3}{2}} < \infty.$$

On the other hand,

$$\int_1^\infty \left| \int_0^{\frac{t}{2}} \frac{g(t-s) - g(t+s)}{s} ds \right| dt$$

$$= \sum_{m=0}^\infty \int_{2^m}^{2^{m+1}} \left| \int_0^{\frac{t}{2}} \frac{g(t-s) - g(t+s)}{s} ds \right| dt$$

$$\geq \sum_{m=2}^\infty \int_{2^m+1}^{2^m+2^{m-1}} \left| \int_0^{\frac{t}{2}} \frac{g(t-s) - g(t+s)}{s} ds \right| dt$$

$$= \sum_{m=2}^\infty \sum_{k=2^m+2}^{2^m+2^{m-1}-1} \int_k^{k+1} \left| \int_0^{\frac{t}{2}} \frac{g(t-s)}{s} ds \right| dt.$$

Since $0 < k - 2^m - 1 < k - 2^m + 1 < \frac{k}{2} \geq \frac{t}{2}$, we have

$$\int_0^{\frac{t}{2}} \frac{g(t-s)}{s}\, ds \geq \int_{k-2^m-1}^{k-2^m+1} \frac{g(t-s)}{s}\, ds$$

$$\geq \frac{1}{k-2^m+1} \int_{k-2^m-1}^{k-2^m+1} g(t-s)\, ds$$

$$= \frac{1}{k-2^m+1} \int_{2^m}^{2^m+1} g(s)\, ds.$$

Consequently,

$$N_1 \geq \sum_{m=2}^{\infty} \frac{1}{(m+1)^2} \sum_{k=2^m+2}^{2^m + 2^{m-1}-1} \frac{1}{k-2^m+1}.$$

The right-hand side is equivalent to

$$\sum_{m=2}^{\infty} \frac{1}{m+1} = \infty,$$

which gives the desired counterexample.

Taking into account Theorem 3.5 and supplementing the above counterexample with the property that $\frac{|f(x)|}{x}$ is integrable, we can construct f with the derivative in $A_{1,2}$ but not in H_Q^1.

For completeness, we give a different example that along with the one above shows that the spaces $A_{1,2}$ and H_o^1 are incomparable. More precisely, we are going to construct a function that belongs to H_o^1 but not to $A_{1,2}$. For this, we recall that in [62] the space H_o^1 enjoys a special atomic characterization. It is equivalent to the usual atomic characterization by means of odd atoms but is more convenient for functions defined on $\mathbb{R}_+$ and then extended as odd functions. The atoms are of two types:

(a) The atoms of first type are of the form $a(t) = \frac{1}{\delta}\chi_{[0,\delta]}(t)$, where $\delta > 0$ and $\chi_{[0,\delta]}$ is as usual the indicator function of the interval $[0, \delta]$.

(b) The atoms of second type are the usual atoms but supported on intervals $[c, d] \subset \mathbb{R}_+$.

It can easily be proved that the atoms of the first type are in $A_{1,2}$. To show that the situation is different for the atoms of the second type, we construct an atom with a very small support located far away from the origin. Let

$$[c, d] = \left[2^N - 1 + \frac{1}{2^{M+1}}, 2^N - 1 + \frac{1}{2^M} \right],$$

with N and M very large integers. We will further specify them later on. We have

$$d - c = \frac{1}{2^{M+1}}.$$

3.3. Intermediate spaces

Let
$$a(c) = -\frac{1}{2^{M+1}}, \quad a(d) = \frac{1}{2^{M+1}},$$

and let $a(t)$ be linear in between. Obviously, a is an atom and thus belongs to H_o^1. It is clear that $m \leq N-1$ should be taken in the $A_{1,2}$ norm. For non-negative m, we have non-zero values only for $0 \leq m \leq N-1$ and $j = 2^{N-m} - 1$. These give the value

$$\sum_{m=0}^{N-1} \int_{2^N-2^m}^{2^N} |a(t)|\,dt = \sum_{m=0}^{N-1} \int_{2^N-1+\frac{1}{2^{M+1}}}^{2^N-1+\frac{1}{2^M}} |a(t)|\,dt = 2 \sum_{m=0}^{N-1} \int_{2^N-1+\frac{3}{2^M}}^{2^N-1+\frac{1}{2^M}} a(t)\,dt$$

$$= 2 \sum_{m=0}^{N-1} \int_{2^N-1+\frac{3}{2^M}}^{2^N-1+\frac{1}{2^M}} (t - 2^N + 1 - \frac{3}{2^M})\,dt$$

$$= \frac{N}{2^{2M+4}}.$$

Obviously, for N very large, say $N = M2^{2M+4}$, when the value is M, the $A_{1,2}$ norms of such atoms cannot be uniformly bounded. These calculations also show that the $A_{1,2}$ norm of an atom with small support will not grow as above if the atom is located close to the origin, for instance, on the interval $[\frac{1}{2^{M+1}}, \frac{1}{2^M}]$.

3.3.3 Intermediate spaces between H_o^1 and H_Q^1

We are now going to find a nested sequence of intermediate spaces between H_o^1 and H_Q^1. They will be applicable to both sine and cosine Fourier transforms. Looking back at the O_q spaces in Subsection 1.4.3, which are the subspaces of H_o^1, we define a symmetric scale but with respect to $B_s g$ rather than for all g.

Definition 3.20. For $1 < q < \infty$, we define

$$\|g\|_{\Sigma_q} = \int_0^\infty \left(\frac{1}{x} \int_{x \leq t \leq 2x} |B_s g(t)|^q dt\right)^{\frac{1}{q}} dx.$$

For the case $q = \infty$,

$$\|g\|_{\Sigma_\infty} = \int_0^\infty \operatorname*{ess\,sup}_{x \leq t \leq 2x} |B_s g(t)|\,dx.$$

Analogously to O_q, these spaces are subspaces of H_Q^1 and are ordered in the way that the smaller q is the wider is the corresponding space. In other words, $\Sigma_{q_1} \subset \Sigma_{q_2}$ provided $q_2 < q_1$.

Proposition 3.21. For $1 < q < 2$, we have

$$H_o^1(\mathbb{R}_+) \hookrightarrow \Sigma_q \hookrightarrow H_Q^1(\mathbb{R}_+).$$

Proof. It has been proved above that for each molecule M in the molecular representation of $g \in H_o^1(\mathbb{R}_+)$ the function $B_s M$ is also a molecule with the norm bounded by a constant, which is the same for all such molecules. Therefore such g also belongs to H_Q^1. To prove the present proposition, it suffices to show that g also belongs to the corresponding Σ_q. In turn, this assertion will be proven if we show that the function

$$\left(\frac{1}{x}\int_{x\leq t\leq 2x}|B_sM(t)|^2 dt\right)^{\frac{1}{2}} dx$$

and this function times x are both square integrable provided that M is a molecule in the molecular decomposition of g. Indeed, all Σ_q with $1 < q < 2$ will then be between Σ_2 and H_Q^1. So, let

$$L_M = \left(\frac{1}{x}\int_{x\leq t\leq 2x}|M(t)|^2 dt\right)^{\frac{1}{2}} dx,$$

where M is a molecule. This is enough since, as is proven above, $B_s M$ is a molecule if M is. We have

$$\int_0^\infty |L_M(x)|^2 dx = \int_0^\infty \frac{1}{x}\int_{x\leq t\leq 2x}|M(t)|^2 dt\, dx = \ln 2 \int_0^\infty |M(t)|^2 dt,$$

which is $\ln 2 \|M\|_2^2$. Further,

$$\int_0^\infty x^2 |L_M(x)|^2 dx \leq \int_0^\infty \frac{1}{x}\int_{x\leq t\leq 2x} t^2 |M(t)|^2 dt\, dx = \ln 2 \int_0^\infty t^2 |M(t)|^2 dt,$$

which is $\ln 2 \|tM(t)\|_2^2$. The proof is complete. $\square$

By this, we have arrived at the following extended chain of embeddings:

$$O_\infty \subset \cdots \subset \underset{1<q<\infty}{O_q} \subset \cdots H_0^1(\mathbb{R}_+) \subset \Sigma_2 \cdots \subset \underset{1<q<2}{\Sigma_q} \subset \cdots$$
$$\subset H_Q^1(\mathbb{R}_+) \subseteq Q_o \subset L_0^1(\mathbb{R}_+).$$

It might be of interest to understand the situation with Σ_q for $q \geq 2$. It is clear that though the spaces Σ_q deliver a scale of intermediate spaces between H_o^1 and H_Q^1, the search for other such spaces, more convenient in applications, is an open problem of considerable importance.

3.4 Fourier–Hardy type inequalities

Back to (1.61), recall that this Fourier–Hardy inequality shows that for the Fourier transform of a function from the Hardy space, we have more than just the Riemann–Lebesgue lemma for an integrable function and its Fourier transform. Recall

3.4. Fourier–Hardy type inequalities

also Proposition 1.18, which is a complement, in a sense, to Corollary 3.4. It is of interest for this section that its formulation may begin in a different manner:

Let the integral on the left-hand side of (1.61) *be finite, or, equivalently, let* $g \in Q$.

The above obtained estimates lead to refinement of (1.61) in certain particular cases. More precisely, for odd functions, (1.61) reduces to

$$\|g\|_{Q_o} \lesssim \|g\|_{H_o^1}.$$

It is proved in Subsection 3.3.2 that this inequality is meaningful, that is, the left-hand side can be finite, with the infinite right-hand side for the same function. Surprisingly, it seems that no such example has been given before in the literature.

A refinement of (1.61) may be an inequality in which one of the following circumstances takes place: either

a) a smaller right-hand side in place of $\|g\|_{H_o^1}$, in other words, for a space wider than H_o^1 but still smaller than Q_o;

or

b) a larger left-hand side in place of $\|g\|_{Q_o}$, that is, for a space smaller than Q_o but still wider than H_0^1;

or

c) both the left-hand side larger than $\|g\|_{Q_o}$ and the right-hand side smaller than $\|g\|_{H_o^1}$.

In fact, we have obtained above the inequalities of all three types. For a), we have

$$\|g\|_{Q_o} \lesssim \|g\|_{H_Q^1} \lesssim \|g\|_{\Sigma_q}.$$

For b), we have

$$\|g\|_{H_Q^1} \lesssim \|g\|_{\Sigma_q} \lesssim \|g\|_{H_o^1}.$$

It is easy to see that in both cases we have double inequalities, since an inequality of type c) holds true:

$$\|g\|_{H_Q^1} \lesssim \|g\|_{\Sigma_q}.$$

Recall that in all the cases $1 < q < 2$.

Accordingly, it is worth mentioning the theorem in [115] or [92, Ch.3] similar to Theorem 2.8 or even more to Theorem 2.28, but for the amalgam type space $A_{1,2}$.

Theorem 3.22. *Let* $f \in BV_0[0, \infty) \cap LAC(0, \infty)$ *and* $f' \in A_{1,2}$. *Then the cosine Fourier transform of* f *is integrable, with*

$$\|\widehat{f_c}\|_{L^1(\mathbb{R}_+)} \lesssim \|f'\|_{A_{1,2}};$$

while for the sine Fourier transform an asymptotic formula holds: for $x > 0$,

$$\widehat{f_s}(x) = \frac{1}{x} f\left(\frac{\pi}{2x}\right) + F(x),$$

where

$$\|F\|_{L^1(\mathbb{R}_+)} \lesssim \|f'\|_{A_{1,2}}.$$

It is also a kind of refinement of the Fourier–Hardy inequality but with the $A_{1,2}$ norm on the right-hand side.

Chapter 4
Sharper results

The results in this chapter have intrinsic interest and importance. But, additionally, this chapter is a sort of departure point or, say, a bridge to further multidimensional generalizations. Most of the earlier obtained results have been generalized to the multivariate case; various theorems of that kind can be found in the survey paper [128]. It is also summarized in [92] and in [122]. All these generalizations fail in sharpness, since too many terms fall into a trap of the remainder values. In other words, one-dimensional results are involved in multivariate extensions mostly as the remainder terms; even their leading terms contribute to the multidimensional leading terms too seldom.

It is easy to understand that in order to get more advanced multivariate generalizations, the asymptotic relations in Theorems 2.8 and 2.20 as well as in Theorem 3.5 should be rewritten in a more precise form. In other words, the remainder terms in these or similar theorems should be controlled in a more explicit manner. However, this is meaningful rather than just philosophy if this can be done in a more or less universal and convenient way. It turns out that the balance operator introduced in Section 1.5 is well adjusted for this. Recall that it is defined (see (1.122)) by means of a generating function φ and takes on an appropriate function g the value

$$B_\varphi g(x) = \frac{1}{x^2} \int_0^\infty g\left(\frac{t}{x}\right) \varphi(t)\, dt.$$

It turns out to be applicable in many situations. For example, for $\varphi(t) = \sin t$, denoting the corresponding operator by B_s, we have

$$\widehat{g_s}(x) = x B_s g(x)$$

and, similarly, $x B_c g(x)$ is the cosine Fourier transform of g. Its only (obvious) property that we will need here is that for $g \in L^1(\mathbb{R}_+)$, we have $B_\varphi \in L^1(\mathbb{R}_+)$

provided
$$\int_0^\infty \frac{|\varphi(t)|}{t}\,dt < \infty,$$
see (1.127) in Corollary 1.126.

One more peculiarity that makes new one-dimensional (and multi-dimensional in what follows) results different from the preceding ones is that, in fact, we do not assume that f' belongs to some Hardy space. This can be done in applications, say, in order to simplify calculations but is not assumed for general relations. What is assumed instead, which is much less restrictive, is that the inverse formula for the Hilbert transform (see, e.g., [100, (4.18) or (5.46)]) of f' holds true almost everywhere. Of course, this is the case if the derivative belongs to the corresponding Hardy space, as in Theorems 2.8 and 2.20. More precisely, we assume either almost everywhere

$$-\mathcal{H}_o(\mathcal{H}_e f')(t) = f'(t) \tag{4.1}$$

or almost everywhere

$$-\mathcal{H}_e(\mathcal{H}_o f')(t) = f'(t). \tag{4.2}$$

Sometimes this might be a good hint which of the two theorems, 4.15 or 4.20, should be used.

We will start with reconsidering Trigub's results on the asymptotic behavior of the Fourier transform of a convex function (see, e.g., [191], [198, 6.4.7]) mentioned in the beginning of Chapter 2. These are very recent results from [127]. In the end of this chapter, we also present Trigub's refinement of one Sz.-Nagy's theorem. More precisely, Sz.-Nagy's sufficient condition is replaced by a more general necessary and sufficient condition. This fits our consideration in all respects.

4.1 The Fourier transform of a convex function

One of the versions of the mentioned Trigub's result can be formulated as follows. We unite the cosine and sine Fourier transforms in one formula by studying, for $\gamma = 0$ or 1,

$$\widehat{f}_\gamma(x) = \int_0^\infty f(t) \cos(xt - \frac{\pi\gamma}{2})\,dt.$$

It is clear that $\widehat{f}_\gamma$ represents the cosine Fourier transform in the case $\gamma = 0$ while taking $\gamma = 1$ gives the sine Fourier transform. Recall that the total variation of f on $[a,b]$ is denoted by $V_{[a,b]}f$.

Theorem 4.3. *Let f be a continuous function defined on $[a,b]$, $0 \le a < b \le +\infty$, convex (upwards or downwards) on $[a,b]$. Then for each $x > 2$,*

$$\widehat{f}_\gamma(x) = \frac{1}{x}f\big(b - \frac{d}{x}\big)\sin\big(bx - \frac{\pi\gamma}{2}\big) - \frac{1}{x}f\big(a + \frac{d}{x}\big)\sin\big(ax - \frac{\pi\gamma}{2}\big) + \theta F(x), \tag{4.4}$$

4.1. The Fourier transform of a convex function

where $d = \min\{\frac{b-a}{2}, \frac{\pi}{2}\}$, F is decreasing on $(2, \infty)$ and such that

$$\int_2^\infty F(x)\,dx \leq \frac{1}{d} V_{[a,b]} f,$$

while $|\theta| \leq C$, with C being an absolute constant.

In fact, the core of much of the work after Trigub was to show that more or less similar asymptotic relations continue to be valid for functions from wider classes, with certain oscillations. Though such general results have numerous applications in the summability of the Fourier series and approximation (e.g., the nice application of such results to approximation theory in [63]), convex functions still remain to be one of the important classes in many problems. First of all, they play an essential role in convex geometry and convex optimization, but also in harmonic analysis they continuously are of importance. Interesting examples can be found in [198, 8.1], where convexity is of crucial importance in the problem of summability of the Fourier series by the method of arithmetic means with gaps.

We first rewrite, in a sense, Theorem 4.3 in a different language, more precisely, to get an explicit expression for $\theta F(x)$, convenient enough and controllable, at least for convex functions. As a consequence of such a precise formula, we will generalize the obtained results to the multidimensional case in Chapter 7.

4.1.1 General representation of the Fourier transform

Here we present a general representation of the Fourier transforms of a function not assuming any really specific conditions. Such a representation starts to work after posing additional assumptions that specify certain class of functions. However, in the presented form the result is well adjusted to convex functions. We will show this in the following subsection by deriving Theorem 4.3 from it. The other feature of such a representation, as mentioned, is its convenience for multidimensional extensions.

Theorem 4.5. *Let f be locally absolutely continuous on (a,b) and let*

$$d = \min\left\{\frac{b-a}{2}, \frac{\pi}{2}\right\}.$$

Then for $x > 0$,

$$\widehat{f}_\gamma(x) = \int_a^b f(t)\cos(xt - \frac{\pi\gamma}{2})\,dt$$
$$= \frac{1}{x}f(b - \frac{d}{x})\sin(bx - \frac{\pi\gamma}{2}) - \frac{1}{x}f(a + \frac{d}{x})\sin(ax - \frac{\pi\gamma}{2}) + B_K f'(x), \quad (4.6)$$

where

$$K(t,x) := K_\gamma(t,x) = \begin{cases} \sin(ax - \frac{\pi\gamma}{2}) - \sin(t - \frac{\pi\gamma}{2}), & ax \leq t < ax + d, \\ -\sin(t - \frac{\pi\gamma}{2}), & ax + d \leq t \leq bx - d, \\ \sin(bx - \frac{\pi\gamma}{2}) - \sin(t - \frac{\pi\gamma}{2}), & bx - d < t \leq bx, \\ 0, & \text{otherwise.} \end{cases}$$

Proof. We have

$$\int_a^{a+\frac{d}{x}} f(t) \cos(xt - \frac{\pi\gamma}{2}) \, dt$$

$$= \int_a^{a+\frac{d}{x}} [f(t) - f(a + \frac{d}{x})] \cos(xt - \frac{\pi\gamma}{2}) \, dt$$

$$+ f(a + \frac{d}{x}) \int_a^{a+\frac{d}{x}} \cos(xt - \frac{\pi\gamma}{2}) \, dt$$

$$= -\int_a^{a+\frac{d}{x}} \cos(xt - \frac{\pi\gamma}{2}) \int_t^{a+\frac{d}{x}} f'(s) \, ds \, dt$$

$$+ f(a + \frac{d}{x}) \int_a^{a+\frac{d}{x}} \cos(xt - \frac{\pi\gamma}{2}) \, dt$$

$$= -\int_a^{a+\frac{d}{x}} f'(s) \int_a^s \cos(xt - \frac{\pi\gamma}{2}) \, dt \, ds$$

$$+ f(a + \frac{d}{x}) \int_a^{a+\frac{d}{x}} \cos(xt - \frac{\pi\gamma}{2}) \, dt$$

$$= \frac{1}{x} \int_a^{a+\frac{d}{x}} f'(s)[\sin(ax - \frac{\pi\gamma}{2}) - \sin(xs - \frac{\pi\gamma}{2})] \, ds$$

$$+ \frac{1}{x} f(a + \frac{d}{x}) \sin(ax + d - \frac{\pi\gamma}{2}) - \frac{1}{x} f(a + \frac{d}{x}) \sin(ax - \frac{\pi\gamma}{2}).$$

Similarly,

$$\int_{b-\frac{d}{x}}^b f(t) \cos(xt - \frac{\pi\gamma}{2}) \, dt = \frac{1}{x} \int_{b-\frac{d}{x}}^b f'(s)[\sin(bx - \frac{\pi\gamma}{2}) - \sin(xs - \frac{\pi\gamma}{2})] \, ds$$

$$- \frac{1}{x} f(b - \frac{d}{x}) \sin(bx + d - \frac{\pi\gamma}{2}) + \frac{1}{x} f(b - \frac{d}{x}) \sin(bx - \frac{\pi\gamma}{2}).$$

4.1. The Fourier transform of a convex function

We then integrate by parts in the rest of the integral:

$$\int_{a+\frac{d}{x}}^{b-\frac{d}{x}} f(t)\cos(xt - \frac{\pi\gamma}{2})\,dt = \frac{1}{x}f(b-\frac{d}{x})\sin(bx+d-\frac{\pi\gamma}{2})$$

$$-\frac{1}{x}f(a+\frac{d}{x})\sin(ax+d-\frac{\pi\gamma}{2}) - \frac{1}{x}\int_{a+\frac{d}{x}}^{b-\frac{d}{x}} f'(t)\sin(xt-\frac{\pi\gamma}{2})\,dt.$$

Combining these three calculations, we obtain

$$\widehat{f_\gamma}(x) = \frac{1}{x}f(b-\frac{d}{x})\sin(bx-\frac{\pi\gamma}{2}) - \frac{1}{x}f(a+\frac{d}{x})\sin(ax-\frac{\pi\gamma}{2})$$

$$-\frac{1}{x}\int_{a+\frac{d}{x}}^{b-\frac{d}{x}} f'(t)\sin(xt-\frac{\pi\gamma}{2})\,dt$$

$$+\frac{1}{x}\int_{a}^{a+\frac{d}{x}} f'(t)[\sin(ax-\frac{\pi\gamma}{2}) - \sin(xt-\frac{\pi\gamma}{2})]\,dt$$

$$+\frac{1}{x}\int_{b-\frac{d}{x}}^{b} f'(t)[\sin(bx-\frac{\pi\gamma}{2}) - \sin(xt-\frac{\pi\gamma}{2})]\,dt. \tag{4.7}$$

To complete the proof, it remains to substitute $xt \to t$ in the integrals of the right-hand side and compare them with the expression for $K(t,x)$. □

This proof might be disappointing. However, (4.6) is the last station after which the (presumably) remainder terms lose either the possibility to make (4.6) to be an explicit equality or the form that makes such an equality "uniform" and controllable. Moreover, in the next subsection such a controllability will be proved for convex functions, while in the second part of the book it will prove its workability for multidimensional generalizations.

The obtained results allow one to easily write down a formula of the above type for the general Fourier transform

$$\widehat{f_\pm}(x) = \int_a^b f(t)e^{\pm ixt}\,dt,$$

where $x > 0$ is taken along with $\pm$ instead of considering x both positive and negative.

Corollary 4.8. *Under assumptions of Theorem 4.5 we have for $x > 0$,*

$$\widehat{f_\pm}(x) = \mp \frac{i}{x}f(b-\frac{d}{x})e^{\pm ibx}$$

$$\pm \frac{i}{x}f(a+\frac{d}{x})e^{\pm iax} + B_{K_g}f'(x), \tag{4.9}$$

where

$$K_g(t,x) = \mp i \begin{cases} e^{\pm iax} + e^{\mp it}, & ax \leq t < ax + d, \\ -e^{\pm it}, & ax + d \leq t \leq bx - d, \\ e^{\pm ibx} + e^{\mp it}, & bx - d < t \leq bx, \\ 0, & \text{otherwise.} \end{cases}$$

The result is obtained by writing Theorem 2.8 for $\gamma = 0$, then for $\gamma = 1$ and summing these two versions, taking into account Euler's formula. We would also like to mention that $a, b \geq 0$ are taken only for simplicity and, in fact, are never used. Such a restriction looks more natural when the cosine and sine Fourier transform are involved but is not very natural in the considered general case.

4.1.2 Convex functions

We will show that for convex functions the last term in (4.6) is L^1 integrable. In other words, we just prove Theorem 4.3, or, more precisely, repeat the main steps of Trigub's proof. First of all, we mention that within (a, b) the function, being convex, is Lipschitz (see, e.g., [151]) and thus locally absolutely continuous. The last two terms in (4.7) are treated in a similar way. We give details for the first one.

$$\int_1^\infty \frac{1}{x} \left| \int_a^{a+\frac{d}{x}} f'(t)[\sin(ax - \frac{\pi\gamma}{2}) - \sin(xt - \frac{\pi\gamma}{2})] \, dt \right| dx$$

$$\leq \int_1^\infty \int_a^{a+\frac{d}{x}} |f'(t)|(t-a) \, dt \, dx$$

$$= \int_a^{a+d} |f'(t)|(t-a) \int_1^{\frac{d}{t-a}} dx \, dt \leq d \int_a^{a+d} |f'(t)| \, dt. \quad (4.10)$$

Similarly,

$$\int_1^\infty \frac{1}{x} \left| \int_{b-\frac{d}{x}}^b f'(t)[\sin(bx - \frac{\pi\gamma}{2}) - \sin(xt - \frac{\pi\gamma}{2})] \, dt \right| dx \leq d \int_{b-d}^b |f'(t)| \, dt.$$

Finally, we integrate by parts in the Stieltjes sense in the remaining integral in (4.7). Take into account that one-sided derivatives of a convex continuous function exist at every point. Moreover, if one restricts oneself to a one-sided derivative (say, left), such a derivative is monotone and admits the Stieltjes integral with respect to df'. This gives

$$-\frac{1}{x} \int_{a+\frac{d}{x}}^{b-\frac{d}{x}} f'(t) \sin(xt - \frac{\pi\gamma}{2}) \, dt = -\frac{1}{x^2} \int_{a+\frac{d}{x}}^{b-\frac{d}{x}} \cos(xt - \frac{\pi\gamma}{2}) \, df'(t)$$

$$= \frac{1}{x^2} f'(b - \frac{d}{x}) \cos(bx - d - \frac{\pi\gamma}{2}) - \frac{1}{x^2} f'(a + \frac{d}{x}) \cos(ax + d - \frac{\pi\gamma}{2}). \quad (4.11)$$

4.1. The Fourier transform of a convex function

In fact, convexity really comes into play when estimating the integral on the right-hand side of (4.11). Indeed, it is dominated by

$$\frac{1}{x^2}\int_{a+\frac{d}{x}}^{b-\frac{d}{x}}|df'(t)|. \tag{4.12}$$

By convexity, df' preserves the sign, and it is equivalent to

$$\frac{1}{x^2}\left|\int_{a+\frac{d}{x}}^{b-\frac{d}{x}} df'(t)\right|. \tag{4.13}$$

The latter leads to the two terms to be estimated,

$$\frac{1}{x^2}|f'(b-\frac{d}{x})|$$

and

$$\frac{1}{x^2}|f'(a+\frac{d}{x})|.$$

The same comes from the last two terms on the right-hand side of (4.11). Since both expressions are estimated in the same manner, let us proceed in detail to one of them, say, the first one. Since f' is monotone, $|f'(b-\frac{d}{x})|$ is the smallest value either on $[b-\frac{d}{x},b-\frac{d}{2x}]$ or on $[b-\frac{2d}{x},b-\frac{d}{x}]$. Hence we have

$$\int_2^\infty \frac{1}{x^2}|f'(b-\frac{d}{x})|\,dx \le \frac{2}{3d}\int_2^\infty \frac{1}{x}\int_{b-\frac{2d}{x}}^{b-\frac{d}{2x}}|f'(t)|\,dt$$

$$= \frac{2}{3d}\int_{b-d}^{b-\frac{d}{4}}|f'(t)|\int_2^{\frac{2d}{b-t}} dx\,dt + \frac{2}{3d}\int_{b-\frac{d}{4}}^{b}|f'(t)|\int_{\frac{d}{2(b-t)}}^{\frac{2d}{b-t}} dx\,dt$$

$$\le \frac{2}{3d}3\ln 2\int_{b-d}^{b}|f'(t)|\,dt = \frac{2\ln 2}{d}\int_{b-d}^{b}|f'(t)|\,dt.$$

In general, we obtain

$$\int_2^\infty |B_K f'(x)|\,dx \le \left(d+\frac{4\ln 2}{d}\right)\int_a^b |f'(t)|\,dt.$$

These estimates also show the monotonicity of the remainder.

Remark 4.14. It is clear that all these results can immediately be extended to functions representable as the difference of those satisfying the corresponding assumptions, in particular, as the difference of two convex functions. The latter class is sometimes called quasi-convex, though there are other classes which bear this name.

4.2 Generalizations of Theorems 2.8 and 2.20

Naturally, we continue with Theorem 2.8, the next after convexity in this type of results. It turns out that it can be expressed in a much more precise form.

Theorem 4.15. *Let $f \in BV_0[0, \infty) \cap LAC(0, \infty)$. Let almost everywhere*

$$-\mathcal{H}_o(\mathcal{H}_e f')(t) = f'(t). \tag{4.16}$$

Then for the cosine Fourier transform of f, we have

$$\widehat{f_c}(x) = -B_s f'(x), \tag{4.17}$$

while for the sine Fourier transform an asymptotic formula holds: for $x > 0$,

$$\widehat{f_s}(x) = \frac{1}{x} f\left(\frac{\pi}{2x}\right) + B_s(\mathcal{H}_o f')(x) + B_S f'(x), \tag{4.18}$$

where B_S is generated by the function

$$S(t) = \frac{2}{\pi} \begin{cases} -t \int_0^\infty \frac{\sin s}{s(s+t)}\, ds, & 0 < t < \frac{\pi}{2}, \\ \int_0^\infty \frac{\sin s}{s+t}\, ds, & t \geq \frac{\pi}{2}. \end{cases}$$

Remark 4.19. This theorem becomes more meaningful if one observes that

$$\|B_s f'\|_{L^1(\mathbb{R}_+)} \lesssim \|f'\|_{H^1_o(\mathbb{R}_+)},$$
$$\|B_s(\mathcal{H}_o f')\|_{L^1(\mathbb{R}_+)} \lesssim \|f'\|_{H^1_o(\mathbb{R}_+)},$$

both by (1.61), provided $f' \in H^1_o(\mathbb{R}_+)$, and

$$\|B_S f'\|_{L^1(\mathbb{R}_+)} \lesssim \|f'\|_{L^1(\mathbb{R}_+)}.$$

Proof. One should accurately keep unestimated all the bounds that give $F(x)$ in the proof of Theorem 2.8. For this, let us recall the main steps of the proof for the sine transform, since for the cosine transform the result follows immediately after integrating by parts. To make use of (1.61), or, more precisely, to separate $B_s(\mathcal{H}_o f')(x)$ as an isolated summand, the inverse formula for the Hilbert transform is applied to the derivative of f after integrating by parts in the sine Fourier transform. This leads to

$$\frac{1}{x} \int_0^\infty (\mathcal{H}_e f')(t) \sin xt\, dt.$$

To replace $\mathcal{H}_e$ by $\mathcal{H}_o$ before applying (1.61), we use (1.59), wherein the above mentioned estimates appear. More precisely, these are (2.8) and calculations after it. Changing variables $xt \to t$ in each of the terms makes this group to become $B_S f'(x)$. That they indeed form an integrable remainder term follows from (1.127). The proof is complete. □

4.2. Generalizations of Theorems 2.8 and 2.20

Like above, Theorem 2.20 can also be expressed in a much more precise form. The details are very similar to those above.

Theorem 4.20. *Let $f \in BV_0[0, \infty) \cap LAC(0, \infty)$. Let almost everywhere*

$$-\mathcal{H}_e(\mathcal{H}_o f')(t) = f'(t). \tag{4.21}$$

Then for the sine Fourier transform of f, we have

$$\widehat{f}_s(x) = \int_0^\infty f(t)\sin xt \, dt = B_c f'(x),$$

while for the cosine Fourier transform an asymptotic formula holds: for $x > 0$,

$$\widehat{f}_c(x) = B_L f'(x) + B_c(\mathcal{H}_e f')(x) + B_C f'(x),$$

where B_L is generated by the function

$$L(t) = \begin{cases} A + \frac{2}{\pi} \ln \frac{2t}{\pi}, & 0 < t < \frac{\pi}{2}, \\ 0, & t \geq \frac{\pi}{2}, \end{cases}$$

and B_C is generated by the function

$$C(t) = -\frac{2}{\pi} \begin{cases} (\cos t - 1)\int_t^\infty \frac{\cos s}{s} ds \\ + \sin t \int_t^\infty \frac{\sin s}{s} ds + \int_0^t \frac{\cos s - 1}{s} ds, & 0 < t < \frac{\pi}{2}, \\ \int_0^\infty \frac{\cos s}{s+t} ds, & t \geq \frac{\pi}{2}. \end{cases}$$

Remark 4.22. This theorem becomes more meaningful if one observes that

$$\|B_c f'\|_{L^1(\mathbb{R}_+)} \lesssim \|f'\|_{H_e^1(\mathbb{R}_+)},$$
$$\|B_c(\mathcal{H}_e f')\|_{L^1(\mathbb{R}_+)} \lesssim \|f'\|_{H_e^1(\mathbb{R}_+)}$$

both by (1.61) provided $f' \in H_e^1(\mathbb{R}_+)$, and

$$\|B_C f'\|_{L^1(\mathbb{R}_+)} \lesssim \|f'\|_{L^1(\mathbb{R}_+)}.$$

Proof. About the leading term first. In Theorem 2.20 we have two leading terms. Here, it is convenient to rewrite exactly the same two terms as one by introducing a special function L. It is possible, since, by absolute continuity and the cancelation property (1.49),

$$\frac{A}{x} f\left(\frac{\pi}{2x}\right) = \frac{A}{x} \int_0^{\frac{\pi}{2x}} f'(t) \, dt.$$

What is unusual here is that even the leading term is expressed by means of the balance operator. Before, we applied it to control the remainder terms.

As above, one should thoroughly follow all the bounds that give $F(x)$ in the proof of Theorem 2.20. We get the ingredients for B_C from the relations after (2.25), from (2.26) and from the calculations after (2.27). □

4.3 The sine Fourier transform revisited

As was mentioned, the study in detail in [121] of asymptotic behavior of the sine Fourier transform of an *arbitrary* function of bounded variation resulted in Theorem 3.5. The reasons were explained why it is not appropriate for multidimensional generalizations. It turns out that accurate analysis of all the estimates in the proof of Theorem 3.5 allows one to rewrite (3.6) in a more advanced form.

Theorem 4.23. *Let $f : \mathbb{R}_+ \to \mathbb{C}$ be locally absolutely continuous on $(0, \infty)$, of bounded variation and $\lim_{t \to \infty} f(t) = 0$. Then for the sine Fourier transform of f there holds, for any $x > 0$,*

$$\widehat{f}_s(x) = \frac{1}{x} f\left(\frac{\pi}{2x}\right) - \mathcal{H}_o B_s f'(x) + B_G f'(x), \qquad (4.24)$$

where B_G is generated by the function

$$G(t) = \begin{cases} G_1(t) = \cos t - 1 - \frac{2}{\pi} \cos t \int_0^t \frac{\sin v}{v} dv \\ \qquad - \frac{2}{\pi} \sin t \int_t^\infty \frac{\cos v}{v} dv, & 0 < t < \frac{\pi}{2}, \\ G_2(t) = -\frac{2}{\pi} \sin t \int_t^\infty \frac{\cos v}{v} dv + \frac{2}{\pi} \cos t \int_t^\infty \frac{\sin v}{v} dv, & t \geq \frac{\pi}{2}, \end{cases} \qquad (4.25)$$

and

$$\|B_G\|_{L^1(\mathbb{R}_+)} \lesssim \|f'\|_{L^1(\mathbb{R}_+)}.$$

Proof. Again, one should accurately keep unestimated all the bounds that give $G(x)$ in the proof of Theorem 3.5. Thus, let us outline how it is obtained.

Starting with integration by parts in

$$\int_0^{\frac{\pi}{2x}} f(t) \sin xt \, dt = \left.\frac{1 - \cos xt}{x} f(t)\right|_0^{\frac{\pi}{2x}} + \frac{1}{x} \int_0^{\frac{\pi}{2x}} f'(t)[\cos xt - 1] \, dt$$

$$= \frac{1}{x} f\left(\frac{\pi}{2x}\right) + \frac{1}{x} \int_0^{\frac{\pi}{2x}} f'(t)[\cos xt - 1] \, dt, \qquad (4.26)$$

we see that the last value is dominated by

$$\int_0^{\frac{\pi}{2x}} t|f'(t)| \, dt,$$

and

$$\int_0^\infty \int_0^{\frac{\pi}{2x}} t|f'(t)| \, dt \, dx = \frac{\pi}{2} \int_0^\infty |f'(t)| \, dt. \qquad (4.27)$$

On the other hand, the last integral in (4.26) can be rewritten as

$$\frac{1}{x^2} \int_0^{\frac{\pi}{2}} f'\left(\frac{t}{x}\right) [\cos t - 1] \, dt,$$

4.3. The sine Fourier transform revisited

which gives $\cos t - 1$ for G_1 in (4.25).

Going over to
$$I = I(x) = \int_{\frac{\pi}{2x}}^{\infty} f(t) \sin xt \, dt,$$

we are going to prove that (see Lemma 3.11 in the previous chapter)

$$-\mathcal{H}_o B_s f'(x) = \frac{2}{x\pi} \int_0^{\infty} f'(t) \sin xt \int_{xt}^{\infty} \frac{\cos v}{v} \, dv \, dt$$
$$+ \frac{2}{x\pi} \int_0^{\infty} f'(t) \cos xt \int_0^{xt} \frac{\sin v}{v} \, dv \, dt. \qquad (4.28)$$

Denoting, in a standard manner,
$$\mathrm{Ci}(u) = -\int_u^{\infty} \frac{\cos t}{t} \, dt$$

and
$$\mathrm{Si}(u) = \int_0^u \frac{\sin t}{t} \, dt = \frac{\pi}{2} - \int_u^{\infty} \frac{\sin t}{t} \, dt,$$

we will make use of the formula (see [15, Ch.II, §2.2 (18)])

$$\int_0^{\infty} \frac{1}{x^2 - u^2} \sin ut \, du = \lim_{\delta \to 0+} \int_{|x-u|>\delta} \frac{1}{x^2 - u^2} \sin ut \, du$$
$$= \frac{1}{x} [\sin xt \, \mathrm{Ci}(xt) - \cos xt \, \mathrm{Si}(xt)], \qquad (4.29)$$

where the integrals are understood in the principal value sense and $x, t > 0$. Since $f' \in L^1(\mathbb{R}_+)$ and the limit in (4.29) is uniform in t, we have

$$\mathcal{H}_o B_s f'(x) = \frac{2}{\pi} \int_0^{\infty} \widehat{f'_s}(u) \frac{1}{x^2 - u^2} \, du$$
$$= \frac{2}{\pi} \int_0^{\infty} f'(t) \int_0^{\infty} \frac{1}{x^2 - u^2} \sin ut \, du \, dt. \qquad (4.30)$$

Applying (3.13) to the inner integral on the right-hand side of (4.30) and using the expressions for Ci and Si completes the proof of (4.28).

With this in hand, we are going to prove that
$$I(x) = -\mathcal{H}_o B_s f'(x) + B_{G_0}(x),$$

where G_0 is G without $\cos t - 1$ in G_1. We also will not forget to make sure that (3.7) holds.

We denote the two summands on the right-hand side of (4.28) by I_1 and I_2. For both, we make use of the fact that

$$\int_{xt}^{\infty} \frac{\cos v}{v} \, dv = O\left(\frac{1}{xt}\right).$$

The same is true if $\cos v$ is replaced by $\sin v$. We begin with I_1. For $t \geq \frac{\pi}{2x}$, we have

$$\int_0^\infty \frac{1}{x} \int_{\frac{\pi}{2x}}^\infty |f'(t)| \frac{1}{xt} \, dt \, dx = \int_0^\infty \frac{|f'(t)|}{t} \int_{\frac{\pi}{2t}}^\infty \frac{1}{x^2} \, dx \, dt$$

$$= \frac{2}{\pi} \int_0^\infty |f'(t)| \, dt. \qquad (4.31)$$

By this, we can rewrite this part of I_1 as

$$\frac{1}{x^2} \int_{\frac{\pi}{2}}^\infty f'\left(\frac{t}{x}\right) \frac{2}{\pi} \sin t \int_t^\infty \frac{\cos v}{v} \, dv \, dt,$$

getting the balance operator for the first summand in G_2, with the opposite sign. For $t \leq \frac{\pi}{2x}$, we split the inner integral in this part of I_1 into two. First,

$$\int_1^\infty \frac{\cos v}{v} \, dv = O(1),$$

and using $\left|\frac{\sin xt}{x}\right| \leq t$, we arrive at the relation similar to (4.27). Further, we have

$$\int_{xt}^1 \left|\frac{\cos v}{v}\right| dv \leq \ln \frac{1}{xt}.$$

By this, integrating in x over $(0, \infty)$, we end up with

$$\int_0^\infty |f'(t)| t \int_0^{\frac{1}{t}} \ln \frac{1}{xt} \, dx \, dt = \int_0^\infty |f'(t)| \, dt.$$

Using, as above, that

$$\int_0^{\frac{1}{t}} \ln \frac{1}{xt} \, dx = \frac{1}{t},$$

we can rewrite this part of I_1 as

$$\frac{1}{x^2} \int_0^{\frac{\pi}{2}} f'\left(\frac{t}{x}\right) \frac{2}{\pi} \sin t \int_t^\infty \frac{\cos v}{v} \, dv \, dt,$$

which generates the balance operator for the last summand in G_1, again with the opposite sign.

Let us proceed to I_2. Using that

$$\frac{1}{x} \left| \int_0^{xt} \frac{\sin v}{v} \, dv \right| = O(t),$$

we arrive, for $t \leq \frac{\pi}{2x}$, at (4.27) and, correspondingly, to the balance operator defined by the intermediate term in G_1. Let now $t \geq \frac{\pi}{2x}$. We have

$$\int_0^{xt} \frac{\sin v}{v} \, dv = \frac{\pi}{2} - \int_{xt}^\infty \frac{\sin v}{v} \, dv.$$

Integrating by parts, we obtain

$$\frac{2}{x\pi} \int_{\frac{\pi}{2x}}^{\infty} f'(t) \cos xt \, dt \frac{\pi}{2} = I.$$

For the integral $\int_{xt}^{\infty} \frac{\sin v}{v} dv$, the estimates are exactly like those in (4.31), which leads to the second summand in G_2.

Combining all these and taking into account the signs, we complete the proof. It remains to observe that in fact all the integrability bounds are the applications of Corollary 1.126 to the appropriate operators. □

Briefly concluding, we now have all the main results for the Fourier transform of a function of bounded variation in the form of precise equalities. Road Clear towards multidimensional extensions.

4.4 A Szökefalvi-Nagy type theorem

Since we depart from Sz.-Nagy's theorem, let us cite it (recall that the Wiener class (algebra) $W_0(\mathbb{R})$ is defined in (1.15)).

Theorem 4.32. *Let an even function* $f \in C_0(\mathbb{R}) \cap LAC$, *while* f' *be locally of bounded variation excluding the points* $0 = t_0 < t_1 < \cdots < t_s$. *If for some* δ *such that* $0 < \delta < N < \infty$, *the integrals*

$$\int_0^\delta t|df'(t)|, \quad \int_N^\infty t|df'(t)|,$$

and

$$\int_{t_k-\delta}^{t_k+\delta} |t-t_k| \ln \frac{1}{|t-t_k|} |df'(t)|, \quad 1 \le k \le s,$$

converge, then $f \in W_0(\mathbb{R})$.

This result was one of the first within the present framework. It always stood off the major line. In a recent paper [197], Sz.-Nagy's result is generalized in such a way that sufficient conditions coincide with the necessary ones. In this subsection, we present Trigub's result in a somewhat extended form (mentioned in [197]), which is the next Theorem 4.35. To present appropriate background, we need additional notation. Recall that we write $h \in L^*(\mathbb{R})$ if h is measurable and satisfies (1.103), where this condition is given in terms of the function f. For any other function, say h, it is written as

$$\|h\|_{L^*} = \int_0^\infty \operatorname*{ess\,sup}_{|t| \ge x} |h(t)| \, dx < \infty.$$

We denote $f \in V_0^*(\mathbb{R})$, if f is locally absolutely continuous on $\mathbb{R} \setminus \{0\}$ (written $LAC(0,\infty)$), $f(\infty) = \lim_{|t|\to+\infty} f(t) = 0$ and

$$\|f\|_{V_0^*} = \|f'\|_{L^*} = \int_0^\infty \operatorname*{ess\,sup}_{|t|\geq x} |f'(t)| dx < \infty.$$

The class $L^*(\mathbb{R})$, applied to the Fourier transform, first appeared in Beurling's paper [23]. Condition $f \in V_0^*$ is less restrictive than convexity but more restrictive than just the finiteness of total variation $f \in BV$. Conditions under which a piecewise convex function or a difference of two convex functions belongs to $W_0(\mathbb{R})$ has been studied by Sz.-Nagy [178] and by Trebels [189] (see also [128, 5.6, 5.7]).

In order to unify the results, we denote by $X(\mathbb{R})$ one of the subspaces of the class of functions of bounded variation we have used in this and previous chapters for obtaining integrability and asymptotic results under assumption that the derivative of the considered function belongs to such a space, say $X = H_o^1$ or O_q. Analogously to the above, we replace $f' \in X(\mathbb{R})$ by the notation $f \in X^*(\mathbb{R})$. Correspondingly, a variety of the proven theorems can be reformulated as follows.

Theorem 4.33. *Let f be a function of bounded variation on $\mathbb{R}$, locally absolutely continuous on $\mathbb{R} \setminus \{0\}$ and such that $\lim_{|t|\to\infty} f(t) = 0$. Let also $f' \in X(\mathbb{R})$, written $f \in X^*(\mathbb{R})$. Then*

$$\int_0^\infty f(x)e^{-ixt}dt = -\frac{i}{x}f\left(\frac{\pi}{2|x|}\right) + F(|x|),$$

with

$$\int_0^\infty F(x)\,dx \leq \|f'\|_{X(\mathbb{R})} := \|f\|_{X^*(\mathbb{R})}.$$

We will also need a "shifted" version of this theorem.

Corollary 4.34. *If $f \in C_0(\mathbb{R})$ and $f \in LAC(\mathbb{R} \setminus \{t_1\})$ for some $t_1 \in \mathbb{R}$, then*

$$\left|\int_{0\leq a\leq |x|\leq b\leq\infty} \left|\widehat{f}(x)\right|\,dx - 2\int_{\frac{\pi}{2b}}^{\frac{\pi}{2a}} \frac{|f(t_1+t) - f(t_1-t)|}{t}dt\right|$$
$$\leq C\|f(\cdot + t_1)\|_{X^*(\mathbb{R})}.$$

Proof. Writing

$$\widehat{f}(x) = \int_\mathbb{R} f(t)e^{-ixt}dt = e^{-it_1 x}\int_\mathbb{R} f(t+t_1)e^{-ixt}dt$$
$$= e^{-it_1 x}\left(\int_0^\infty f(t+t_1)e^{-ixt}dt + \int_0^\infty f(t_1-t)e^{ixt}dt\right),$$

we have, by Theorem 4.33,

$$\widehat{f}(x) = -\frac{i}{y}e^{-it_1 x}\left(f\left(t_1 + \frac{\pi}{2|x|}\right) - f\left(t_1 - \frac{\pi}{2|x|}\right)\right) + F_1(|x|),$$

with

$$\int_0^\infty |F_1(x)|\, dx \leq \|f(\cdot + t_1)\|_{X^*(\mathbb{R})}.$$

This yields

$$\left||\widehat{f}(x)| - \frac{1}{|x|}\left|f\left(t_1 + \frac{\pi}{2|x|}\right) - f\left(t_1 - \frac{\pi}{2|x|}\right)\right|\right| \leq C|F_1(|x|)|$$

and

$$\left|\int_{a\leq |x|\leq b} |\widehat{f}(x)|\, dx - \int_{a\leq |x|\leq b} \frac{|f\left(t_1 + \frac{\pi}{2|x|}\right) - f\left(t_1 - \frac{\pi}{2|x|}\right)|}{|x|}\, dx\right|$$
$$\leq 2C\|f(\cdot + t_1)\|_{X^*(\mathbb{R})}.$$

It remains to take into account that the last integral on the left-hand side is just

$$2\int_{\frac{\pi}{2b}}^{\frac{\pi}{2a}} \frac{|f(t_1 + t) - f(t_1 - t)|}{t}\, dt,$$

which completes the proof. $\square$

We are now in a position to formulate and prove the mentioned generalization of Sz.-Nagy's theorem.

Theorem 4.35. *Let $\{t_k\}_{k=1}^s$ be the points on $\mathbb{R}$ such that the functions $f(t + t_k)$, $1 \leq k \leq s$, admit extensions from a neighborhood of the origin and f itself from a neighborhood of infinity ($|t| \geq N$) to functions from Theorem 4.33 with derivatives in $X(\mathbb{R})$. In order that $f \in W_0(\mathbb{R})$ it is necessary and sufficient, for some $\delta > 0$ and $N > \max_{1\leq k\leq s} |t_k|$,*

$$\sum_{k=1}^s \int_0^\delta \frac{|f(t_k + t) - f(t_k - t)|}{t}\, dt + \int_N^\infty \frac{|f(t) - f(-t)|}{t}\, dt < \infty$$

to hold.

Proof. Recall that if $f \in C_0(\mathbb{R}) \cap BV(\mathbb{R})$, then for each $x \in \mathbb{R}$ (see, e.g., [198, 3.3.10])

$$f(t) = \frac{1}{2\pi} \lim_{\varepsilon_1 \to +0} \lim_{\varepsilon_2 \to +0} \int_{\varepsilon_1 \leq |x| \leq \frac{1}{\varepsilon_2}} \widehat{f}(x) e^{itx}\, dx.$$

Therefore, the proof reduces to $\widehat{f} \in L^1(\mathbb{R})$ to be checked.

Set
$$\delta = \frac{1}{3} \min_{k \neq m} |t_k - t_m|,$$
keeping in mind that each of the functions $f(t + t_k)$ can be extended from the neighborhood of the origin of radius 2δ to a function in $X^*(\mathbb{R})$.

For $|t - t_k| \leq \delta$ and $|t - t_k| > \delta$, set, correspondingly,
$$h_k(t) = 1 \quad \text{and} \quad h_k(t) = \left(2 - \frac{1}{\delta}(t - t_k)\right)_+, \quad 1 \leq k \leq s.$$

Since $h_k \in W_0(\mathbb{R}) \cap X^*(\mathbb{R})$, as a function with compact support in Lip 1, the function $f_k(t) = h_k(t)f(t) \in X^*(\mathbb{R})$ as well (see Lemma 7.19 in Chapter 7 where a corresponding result is proven for any dimension).

Further, by Corollary 4.34, we have that $f_k \in W_0(\mathbb{R})$ if and only if
$$\int_0^\delta \frac{|f_k(t_k + t) - f_k(t_k - t)|}{t} dt < \infty.$$

But $f_k = f$ if $t \in [0, \delta]$, since $h_k(t_k + t) = 1$. By this we have proved the necessity of the conditions in neighborhoods of the points $\{x_k\}, 1 \leq k \leq s$ indicated in the theorem.

Let us now proceed to the sufficiency. We represent f as
$$f(t) = \sum_{k=1}^s f_k(t) + f_{s+1}(t).$$

Here $f_{s+1}(x) = f(x)$ if $|x| \geq N = 2\delta + \max_k |x_k|$, while for $|x| \leq N - \delta$ we have $f_{s+1} \in \text{Lip 1}$. For $|x| \geq N$ and $|x| < N$, set respectively $h_{s+1}(x) = 1$ and
$$h_{s+1}(x) = \frac{1}{\delta}(|x| - N + \delta)_+.$$

Then
$$f_{s+1}(x) = f_{s+1}(x)h_{s+1}(x) + f_{s+1}(x)(1 - h_{s+1}(x)).$$

The second summand, as a function in Lip 1, belongs to $W_0(\mathbb{R}) \cap V^*(\mathbb{R})$. The first summand, by Corollary 4.34, belongs to $W_0(\mathbb{R})$ if and only if
$$\int_{N-\delta}^\infty \frac{|f_{s+1}(x)h_{s+1}(x) - f_{s+1}(-x)h_{s+1}(-x)|}{x} dx < \infty,$$
or, which is the same,
$$\int_N^\infty \frac{|f(x) - f(-x)|}{x} dx < \infty.$$

By this, both the necessity and sufficiency are proved. $\square$

Part II:
Multi-dimensional Case

The previous part was purely one-dimensional. This part will be mostly multi-dimensional but with one-dimensional touches. The point is that some one-dimensional prototypes are so tied with their extensions to any dimension that it seems logical to put them together. Nevertheless, this part is first of all multidimensional, "it's laid for a great many more than three" as Lewis Carroll's Alice said once. Many of the obtained results appear in dimension two to be more or less exactly as in any dimension, it is just the question of notation. However, there are results that in dimension two or three appear different, moreover, a result is given where only the four-dimensional case covers all possible situations.

Chapter 5
A toolkit for several dimensions

As in the previous part, we start with certain preliminaries. There will be similar sections, like those on the Fourier transform or Hardy spaces. However, even in such cases there will be phenomena and properties inherent in the multidimensional case. And, of course, there will be topics one can face only in the case of several variables. The well-known and very often quoted

Oh, the little more, and how much it is!
And the little less, and what worlds away!

by Robert Browning, is suitable for this situation as well.

Before proceeding to specific topics we, naturally, begin with specific notation.

5.1 Indicator notation

As is generally accepted, the success of many of the multidimensional results (more precisely, clarity of their formulations and likeness of the proof to that in dimension one) strongly depends on appropriate notation. We suggest universal indicator type notation that easily allows one to distinct different phenomena on certain groups of variables and in many cases minimize the number of indices. One of the hints that such a notation may be helpful can be found in [146, Ch. 4 and 5]. The main tool for this are zero-one vectors η, χ and ζ. The first one will be constantly used, while the other two only in the situations where there are more than two different phenomena for one function. Thus, the vector η will always be used in the sense described below, and the vectors χ and ζ will always have the same meaning.

Let $\eta = (\eta_1, \ldots, \eta_n)$ be an n-dimensional vector with the entries either 0 or 1 only. Its main task is to indicate the variable in which a certain action be fulfilled. Correspondingly, $|\eta| = \eta_1 + \cdots + \eta_n$. The inequality of vectors is meant coordinate wise. If the only 1 entry is on the j-th place, while the rest are zeros, such a (basis) vector will be denoted by e_j. By x_η we denote the $|\eta|$-tuple consisting only of x_j such that $\eta_j = 1$ and

$$dx_\eta := \prod_{j:\eta_j=1} dx_j.$$

Since we are going to deal with multidimensional variations, various differences will be involved. Denote by $\Delta_{u_\eta} f(x)$ the partial difference

$$\Delta_{u_\eta} f(x) = \left(\prod_{j:\eta_j=1} \Delta_{u_j} \right) f(x),$$

with

$$\Delta_{u_j} f(x) = f(x + u_j e_j) - f(x).$$

We shall freely use $\prod$ for both usual multiplication and repeated operator action. It will be clear each time what is meant and hopefully this will cause no confusion.

Here and in what follows $D^\eta f$ for $\eta = \mathbf{0} = (0, 0, \ldots, 0)$ means $D^\mathbf{0} f := f$, that is, the function itself, while $\eta = \mathbf{1} = (1, 1, \ldots, 1)$ means the partial derivative $D^\mathbf{1} f$ of order n applied repeatedly in each variable, where

$$D^\eta f(x) = \left(\prod_{j:\eta_j=1} \frac{\partial}{\partial x_j} \right) f(x)$$

means the partial derivative of order $|\eta|$ applied to the variables indicated by η. We shall naturally denote D^{e_j} by D^j.

Certain additional notation is in order. As usual, for any vector

$$\alpha = (\alpha_1, \ldots, \alpha_n),$$

we denote

$$x^\alpha = x_1^{\alpha_1} \cdots x_n^{\alpha_n}.$$

Analogously, if $\alpha \in \mathbb{Z}_+^n$, then $D^\alpha f$ denotes the partial derivative of f of order $\alpha_1 + \cdots + \alpha_n$, that is, of order α_j with respect to x_j, $j = 1 \ldots, n$.

To denote the $|\eta|$-tuple consisting only of $\frac{1}{x_j}$ for j such that $\eta_j = 1$, we incorporate negative indicator vectors, that is, the described $|\eta|$-tuple will be denoted by $x_{-\eta}$. Correspondingly, the vector $(\frac{1}{x_1}, \ldots, \frac{1}{x_n})$ will be denoted by $x_{-\mathbf{1}}$.

If in the multivariate setting one of the operators like $\mathcal{H}_o$, B_s, B_G is applied to the j-th variable, it will be denoted by $\mathcal{H}_o^j$, B_s^j, B_G^j, etc. Like the derivative above, the other operators applied to the j-th variables for j such that $\eta_j = 1$ will be denoted by means of the superscript η, i.e.,

$$\mathcal{H}_o^\eta = \prod_{j:\eta_j=1} \mathcal{H}_o^j, \quad B_s^\eta = \prod_{j:\eta_j=1} B_s^j, \quad B_G^\eta = \prod_{j:\eta_j=1} B_G^j.$$

Recall that by B_φ we denote the balance operator introduced in Section 1.5. Its structure obviously allows one to apply it separately to the desired variables.

When we apply the general Hilbert transform, there is no need for superscripts. Such a transform applied to the j-th variable will be defined by $\mathcal{H}_j$ and, consequently, $\mathcal{H}_j \mathcal{H}_k \cdots \mathcal{H}_l := \mathcal{H}_{jk\ldots l}$. For the latter case, the introduced indicator notation is more convenient. Naturally,

$$\mathcal{H}_\eta := \prod_{j:\eta_j=1} \mathcal{H}_j.$$

Let
$$[a,b] = [a_1, b_1] \times [a_2, b_2] \times \cdots \times [a_n, b_n]$$
denote an n-dimensional parallelepiped. Some of the a_j can be $-\infty$, while some of b_j can be $+\infty$.

5.2 Multidimensional variations

There is a variety of notions of bounded variation in several dimensions (besides the old sources [43] and [3], see a very recent paper [32]). The main variation used in generalizations of the obtained one-dimensional results will be Hardy's variation, which is, in turn, a restriction of Vitali's variation. We shall also touch on Tonelli's variation. In Chapter 1, a list of the properties of functions with bounded variation is given. The reasons that there are many variations in several dimensions is not only the greater number of degrees of freedom in the multivariate case but also the desire to generalize as many properties of one-dimensional variation as possible. However, each time one would sacrifice some of them, which depends on the problem the corresponding variation is used in. In any case, the mentioned variations are the most popular.

5.2.1 Vitali's and Hardy's variations

One of the simplest and direct generalizations of the one-dimensional variation, the Vitali variation, is defined as follows (cf., e.g., [43, 3]). Let f be a complex-valued function and

$$\Delta_u f(x) = \Delta_{u_1} f(x) = \left(\prod_{j=1}^{n} \Delta_{u_j}\right) f(x)$$

be a "mixed" difference with respect to the parallelepiped

$$[x, x+u] = [x_1, x_1 + u_1] \times \cdots \times [x_n, x_n + u_n].$$

Let us take an arbitrary number of non-overlapping parallelepipeds, and form the mixed difference with respect to each of them. Then the Vitali variation is

$$VV(f) = \sup \sum |\Delta_u f(x)|,$$

where the sum and then the least upper bound are taken over all the sets of such nonoverlapping parallelepipeds. For smooth functions f (say, absolutely continuous), the Vitali variation is expressed as the following integral

$$VV(f) = \int_{\mathbb{R}^n} \left| \frac{\partial^n f(x)}{\partial x_1 \cdots \partial x_n} \right| dx = \int_{\mathbb{R}^n} |D^1 f(x)| \, dx. \qquad (5.1)$$

Generally,

$$VV(f) = \int_{\mathbb{R}^n} |d^1 f(x)|, \qquad (5.2)$$

where $d^1 f$ denotes the Stieltjes integration with respect to the function of bounded variation f.

Even this variation has numerous applications; besides those in [3], see also [198, Sec.3.3.9-3.3.10]. However, in many problems Vitali's variation is powerless, because marginal functions of a smaller number of variables may be added to a function of bounded Vitali's variation, and such functions can be very bad. The next notion does not have this disadvantage.

Definition 5.3. A function f is said to be of bounded *Hardy variation*, written $f \in VH(f)$, if it is of bounded Vitali variation (in all the variables) and is of bounded Vitali variation with respect to any smaller number of variables.

In fact, the Vitali variation coincides with the usual bounded variation when it is taken with respect to a single variable; see, e.g., [79], [43], [3]. Sometimes the notion in Definition 5.3 is also attributed to Krause, see, e.g., [89, p. 345]). We would like to mention that even in the first publication [79], the main properties of this variation were given in detail.

The fact of being of bounded Vitali variation with respect to a group of variables will be denoted by $VV_\eta(f) < \infty$, with $\eta \neq \mathbf{1}, \mathbf{0}$. Correspondingly, $VV(f) := VV_\mathbf{1}(f)$. In other words, $VH(f) < \infty$ if and only if $VV_\eta(f) < \infty$ for all η, except $\eta = \mathbf{0}$ which is meaningless. However, just for convenience, we can understand $VV_\mathbf{0}(f) := f$. Similarly to (5.2),

$$VV_\eta(f)(x_{1-\eta}) = \int_{\mathbb{R}^{|\eta|}} |d^\eta f(x_\eta, x_{1-\eta})|, \qquad (5.4)$$

where $d^\eta f$ denotes the Stieltjes integration with respect to the corresponding variables.

To see the difference between the two variations, let us mention the analogs of the property (6) in Section 1.1. In fact, each time a new variation was invented, the form of that property, or in other words, analogs of monotonicity were searched for. Thus, see [3], the necessary and sufficient condition that f be of bounded Vitali's variation is that it be expressible as the difference between two functions, f_1 and f_2, satisfying the inequalities $\Delta_{u_1} f_j(x) \geq 0$, $j = 1, 2$. On the other hand, the necessary and sufficient condition that f be of bounded Hardy's variation is that it be expressible as the difference between two functions, f_1 and f_2, satisfying the inequalities $\Delta_{u_\eta} f_j(x) \geq 0$, $j = 1, 2$, for all $\eta \neq \mathbf{0}$.

If f is of bounded Vitali variation on $\mathbb{R}^n$ and $\lim_{|x| \to \infty} f(x) = 0$, then functions depending on a smaller number of variables than n are excluded. Such a function is of bounded Hardy variation. To the best of our knowledge, it was Trigub who paid attention to this interesting feature.

It is worth mentioning that one of the reasons why the results in Part I are mainly generalized to functions with bounded Hardy variation is the fact that this way is not only natural but is very convenient, since allows us to continue to use the Hilbert transform and its properties as a basic machinery.

5.2.2 Tonelli's variation

Also, the so-called Tonelli variation (written $f \in VT$, see [187]) will be a focus of our attention. A function f is of bounded Tonelli variation if for almost every

$$(x_1, \ldots, x_{j-1}, x_{j+1}, \ldots, x_n) = x_{1-\mathbf{e_j}}$$

it is of bounded variation in the single variable x_j for all $1 \leq j \leq n$ and if these variations

$$V_j(f) := V_j^f(x_1, \ldots, x_{j-1}, x_{j+1}, \ldots, x_n)$$

are Lebesgue integrable as functions of the other $n-1$ variables:

$$VT(f) = \sum_{j=1}^{n} \int_{\mathbb{R}^{n-1}} V_j^f(x_1, \ldots, x_{j-1}, x_{j+1}, \ldots, x_n) \prod_{\substack{k=1, \\ k \neq j}}^{n} dx_k$$

$$= \sum_{j=1}^{n} \int_{\mathbb{R}^{n-1}} V_j^f(x_{1-\mathbf{e_j}}) \, dx_{1-\mathbf{e_j}}.$$

For a smooth enough function f, it is equal to

$$VT(f) = \int_{\mathbb{R}^n} \sum_{j=1}^{n} \left| \frac{\partial f(x)}{\partial x_j} \right| dx. \qquad (5.5)$$

Among the sources dealing with the Tonelli variation, let us mention [37] and more recent books [72], [209], [7] and papers [44], [33]. However, these and many other authors deal with such a variation in a more general setting, say, the one where the derivatives are understood in a generalized sense. In our study, we do not need such a generality, since we mainly deal with absolutely continuous functions.

5.3 Fourier transform

We are forced to repeat some of the basics for the Fourier transform from the corresponding place of Part I. As promised, they will be given in a more regular way than in dimension one. In addition, notation and some peculiarities of the multivariate case need to be mentioned and explained. We follow the minimal model in [92, Ch.1].

5.3.1 L^1-theory

Let $f \in L^1(\mathbb{R}^n)$. Define the Fourier transform of f by the formula

$$\widehat{f}(x) = \int_{\mathbb{R}^n} f(u) e^{-i\langle x, u \rangle} du, \tag{5.6}$$

where

$$\langle x, u \rangle = x_1 u_1 + \cdots + x_n u_n.$$

Theorem 5.7. *The mapping $f \to \widehat{f}$ is a bounded linear map from $L^1(\mathbb{R}^n)$ to $L^\infty(\mathbb{R}^n)$, $\widehat{f}$ is continuous, $\lim_{|x| \to \infty} \widehat{f}(x) = 0$, and*

$$\|\widehat{f}\|_{L^\infty(\mathbb{R}^n)} \leq \|f\|_{L^1(\mathbb{R}^n)}. \tag{5.8}$$

The first two properties follow directly from the definition of the Fourier transform. In order to prove the third claim, known as the Riemann–Lebesgue lemma, one should first prove it for the characteristic (indicator) function of a cube by an explicit calculation, and then apply a limiting argument as in Chapter 1 (see, e.g., [99]). The estimate (5.8) follows by the definition of the Fourier transform.

Theorem 5.9. *Let $t_a f(x) = f(x - a)$. Then*

$$\widehat{t_a f}(x) = e^{i\langle a, x \rangle} \widehat{f}(x).$$

Similarly, the Fourier transform of $e^{i\langle a, u \rangle} f(u)$ is $\widehat{f}(x - a)$.

With this simple example, it is reasonable to ask how the Fourier transform behaves under the influence of general linear transformations.

Theorem 5.10. *Let T be a non-singular complex valued linear map from $\mathbb{R}^n$ to itself and define $f_T(u) = f(T(u))$. Then*

$$\widehat{f_T}(x) = |\det T|^{-1} \widehat{f}((T^{-1})^*(x)),$$

where T^* denotes the adjoint of T and $\det T$ is the determinant.

In particular, Theorem 5.10 gives that if $f(u)$ is radial (depending only on $|u|$), then $\widehat{f}$ is radial too. More precisely, it is represented as a Hankel transform as follows.

Theorem 5.11. *Let $t^{n-1} f_0(t)$ be integrable on $(0, +\infty)$. Then for its radial extension $f(x) = f_0(|x|)$,*

$$\widehat{f}(x) = (2\pi)^{\frac{n}{2}} \int_0^{+\infty} f_0(t)(|x|t)^{1-\frac{n}{2}} J_{\frac{n}{2}-1}(|x|t) t^{n-1} dt. \tag{5.12}$$

5.3. Fourier transform

This formula is referred to in [27] as the Cauchy–Poisson formula. Here and in the sequel, J_ν is the Bessel function of first kind and order ν.

We now turn towards the issue of the behavior of the Fourier transform under the influence of differentiation.

Theorem 5.13. *Suppose that $D^j f \in L^1(\mathbb{R}^n)$. Then*

$$\widehat{D_j f}(x) = i x_j \widehat{f}(x).$$

Similarly, if $u_j f(u) \in L^1(\mathbb{R}^n)$, then

$$-i \widehat{x_j f}(x) = D^j \widehat{f}(x).$$

Both identities follow readily from the definition using integration by parts. They can easily be extended to the more general $D^\eta f$ and $u^\eta f$.

As in dimension one, not forgetting that we are mainly interested in conditions for the integrability of the Fourier transform, we also face a related problem of whether a uniformly continuous and vanishing at infinity function f can be representable as the Fourier integral of an integrable function g, written

$$f(u) = \int_{\mathbb{R}^n} g(x) e^{i\langle x, u\rangle} dx. \tag{5.14}$$

It is said (cf. (1.15)) in this case that f belongs to the Wiener space (algebra) $W_0(\mathbb{R}^n)$. Of course, in many situations g can be understood, in one sense or another, as the Fourier transform of f and the last formula as the Fourier inversion. A comprehensive overview of these problems is given in [128].

5.3.2 L^2- and L^p-theory

While this may appear slightly paradoxical, we begin the section on L^2 theory by defining a set of functions which are rather more regular.

Definition 5.15. We say that $\phi \in C^\infty(\mathbb{R}^n)$ belongs to $\mathcal{S}(\mathbb{R}^n)$ if

$$\sup_{x \in \mathbb{R}^n} |x^\gamma D^\alpha \phi(x)| < \infty$$

for all multi-indices γ and α.

It follows easily from Theorem 5.13 that the Fourier transform maps $\mathcal{S}(\mathbb{R}^n)$ to itself. Moreover, we have the following fundamental fact.

Theorem 5.16. *The Fourier transform is an isomorphism of $\mathcal{S}(\mathbb{R}^n)$ into itself whose inverse is given by the formula*

$$f(u) = (2\pi)^{-n} \int_{\mathbb{R}^n} \widehat{f}(x) e^{i\langle x, u\rangle} dx.$$

To see this, we need the following basic calculations.

Lemma 5.17. *Let $f, g \in L^1(\mathbb{R}^n)$. Then*
$$\int_{\mathbb{R}^d} \widehat{f}(x) g(x) \, dx = \int_{\mathbb{R}^n} f(x) \widehat{g}(x) \, dx.$$

The proof of this is immediate by Fubini. We also need to know that, roughly speaking, the Fourier transform of a Gaussian is a Gaussian.

Lemma 5.18. *Let $\gamma(u) = e^{-\frac{i|u|^2}{2}}$. Then*
$$\widehat{\gamma}(x) = (2\pi)^n \gamma(2\pi x).$$

The proof is by completing the square and changing the contour of integration.

Using Lemma 5.18, it is not difficult to derive at the following basic relation, known as the Fourier inversion formula.

Theorem 5.19. *Suppose that $f \in L^1$, and assume that $\widehat{f}$ is also in L^1. Then for almost every x,*
$$f(u) = \int_{\mathbb{R}^n} e^{i\langle x,u \rangle} \widehat{f}(x) \, dx.$$

Perhaps the most fundamental result in Fourier analysis is Plancherel's theorem.

Theorem 5.20. *If $u, v \in \mathcal{S}$, then*
$$\int_{\mathbb{R}^n} \widehat{u}(x) \overline{\widehat{v}}(x) \, dx = \int_{\mathbb{R}^n} u(x) \overline{v}(x) \, dx.$$

Moreover, there is a unique operator $\mathcal{F} : L^2(\mathbb{R}^n) \to L^2(\mathbb{R}^n)$ such that $\mathcal{F}f = \widehat{f}$ when $f \in \mathcal{S}(\mathbb{R}^n)$. This operator is unitary and $\mathcal{F}f = \widehat{f}$ when $f \in L^1 \cap L^2$.

For functions $f \in L^p(\mathbb{R}^n)$, with $1 \leq p \leq 2$, the above theory extends by means of the Hausdorff–Young inequality. It reads, with $\frac{1}{p'} = 1 - \frac{1}{p}$, as

$$\|\widehat{f}\|_{L^{p'}(\mathbb{R}^n)} \leq \|f\|_{L^p(\mathbb{R}^n)}. \tag{5.21}$$

What is said on the sharpness of (5.21) in dimension one in Chapter 1, completely fits the case of several dimensions.

5.3.3 Poisson summation formula

The Poisson summation formula was discovered by Siméon Denis Poisson and is sometimes called Poisson resummation. A typical form of the Poisson summation formula for integrable functions is given in [176, Ch.VII, Th.2.4]. Here, for $\mathbb{T} = (-\pi, \pi]$, we denote $\mathbb{T}^n = \mathbb{T} \times \cdots \times \mathbb{T}$.

Theorem 5.22. *Suppose $f \in L^1(\mathbb{R}^n)$. Then the series*

$$\sum_{m \in \mathbb{Z}^n} f(x+m) \qquad (5.23)$$

converges in the $L^1(\mathbb{T}^n)$ norm. The resulting function in $L^1(\mathbb{T}^n)$ has the Fourier expansion

$$(2\pi)^{-n} \sum_{m \in \mathbb{Z}^n} \widehat{f}(m) e^{i\langle x, m\rangle}.$$

This means that $\{(2\pi)^{-n} \widehat{f}(m)\}$ is the sequence of the Fourier coefficients of the L^1 function defined by the series (5.23), *where, for any $x \in \mathbb{R}^n$, we have* (5.6).

Under certain restrictions (see, e.g., [176, Ch.VII, Cor.2.6]), one has

$$\sum_{m \in \mathbb{Z}^n} f(x+m) = (2\pi)^{-n} \sum_{m \in \mathbb{Z}^n} \widehat{f}(m) e^{i\langle x, m\rangle},$$

and, in particular,

$$\sum_{m \in \mathbb{Z}^n} f(m) = (2\pi)^{-n} \sum_{m \in \mathbb{Z}^n} \widehat{f}(m).$$

Moreover, results are known which show that the Poisson summation characterizes, in that sense or another, the Fourier transform (see [46] and [52]).

Good sources for various versions of the Poisson summation formula and their applications are [142, Ch.X, §6] and [21].

5.4 Multidimensional spaces

Hardy's variation, at least in our study, is closely related with the product Hardy spaces $H_m^1 = H^1(\mathbb{R}^{n_1} \times \cdots \times \mathbb{R}^{n_m})$. These spaces are of interest and importance in certain questions of Fourier Analysis (see, e.g., [40], [53], [71]). Roughly speaking, they appear when all the n variables are split in groups and the function is in the real Hardy space with respect to each of the group and every combination of groups still claims for certain Hardy space conditions with respect to the rest of the variables. A natural question arises, how many such spaces exist for the fixed dimension. It is clear that if the case $m=1$ is not taken into account, which we cannot call product and which is classical real Hardy space $H^1(\mathbb{R}^n$, the only case with one option is that where $m=n$ and, correspondingly, $n_1 = \cdots = n_n = 1$. In all other cases there are plenty of product Hardy spaces. Denoting its number by $S(n,m)$, one can easily find out that

$$S(n,2) = 2^{n-1} - 1$$

and

$$S(n, n-1) = \frac{n(n-1)}{2}.$$

For the rest of $S(n,m)$ calculations are not that simple and turn out to be the classical Stirling set numbers (or the Stirling numbers of the second kind), see, e.g., [76] or more recent [143]. For every $m = 2, 3, \ldots, n-1$, we have

$$S(n,m) = \frac{1}{m!} \sum_{k=0}^{m} (-1)^{m-k} \binom{m}{k} k^n.$$

Of course, the cases $m = 1$ and $m = n$ are subject to this formula as well. It is worth mentioning that knowing these numbers for dimension n, one can find them for the next dimension by means of the relation

$$S(n+1, m) = mS(n, m) + S(n, m-1).$$

Finally, to know the number of all possible product Hardy spaces on $\mathbb{R}^n$, one has to sum up $S(n,m)$ in m:

$$\sum_{m=2}^{n} S(n,m) = \sum_{m=2}^{n} \frac{1}{m!} \sum_{k=0}^{m} (-1)^{m-k} \binom{m}{k} k^n$$
$$= \sum_{k=1}^{n} (-1)^k k^n \sum_{m=k}^{n} \frac{(-1)^m}{m!} \binom{m}{k} - 1.$$

We know from the dimension one computations and estimates that most of the considered proofs begin with integration by parts. The way we are going to integrate by parts, or, correspondingly, differentiate the function in accordance of their belonging to the class of functions with bounded Hardy's variation, lead us to a special case of these spaces where $m = n$ and, correspondingly, $n_1 = \cdots = n_n = 1$. In this situation, we can use the notation

$$H^1(\mathbb{R} \times \cdots \times \mathbb{R}) := H^1_n(\mathbb{R} \times \cdots \times \mathbb{R})$$

and a modified definition on the basis of the Hilbert transforms applied to each variable as in [69].

Since we shall mainly deal with Hardy spaces of even or odd functions in certain variables, correspondingly defined on $\mathbb{R}^n_+$, we need more delicate notation for this. We shall define $H^1_\eta(\mathbb{R} \times \cdots \times \mathbb{R})$ if the Hilbert transform taken with respect to each of the j-th variables, for which $\eta_j = 1$, is the even Hilbert transform, while for the rest it is odd. We will be especially interested in the case where $\eta = \mathbf{0}$, that is, all the Hilbert transforms are odd.

By this, in the general case the norm of a function $g(x)$ in $H^1(\mathbb{R} \times \cdots \times \mathbb{R})$ will be

$$\|g\|_{H^1(\mathbb{R} \times \cdots \times \mathbb{R})} = \sum_{\mathbf{0} \leq \eta \leq \mathbf{1}} \|\mathcal{H}_\eta g\|_{L^1(\mathbb{R}^n)}. \qquad (5.24)$$

5.4. Multidimensional spaces

Of course, the case of $H^1_\eta(\mathbb{R} \times \cdots \times \mathbb{R})$ is subject to corresponding alterations. More precisely,

$$\|g\|_{H^1_\eta(\mathbb{R}\times\cdots\times\mathbb{R})} = \sum_{0\leq\chi\leq 1} \left\| \left(\prod_{j:\chi_j=\eta_j=1} \mathcal{H}^j_e \right) \left(\prod_{j:\chi_j=1,\eta_j=0} \mathcal{H}^j_o \right) g \right\|_{L^1(\mathbb{R}^n)}. \quad (5.25)$$

For this space, the one-dimensional Fourier-Hardy inequality (1.61) is naturally changed to

$$\int_{\mathbb{R}^n} \frac{|\widehat{g}(x)|}{|x_1|\cdots|x_n|} \, dx \lesssim \|g\|_{H^1(\mathbb{R}\times\cdots\times\mathbb{R})}. \quad (5.26)$$

It is given in [71] in dimension 2, with the reference to [95] for the proof. However, in [95] the result is also two-dimensional, which may lead to certain misunderstandings for higher dimensions. A simple inductive argument proves (5.26) in full generality.

Proof. Indeed, using (1.61) as a basis of induction and assuming (5.26) to hold in dimension $n-1$, let us prove it for dimension n. Denoting, like above for other transforms, F_η to be the Fourier transform with respect to the variables x_j for which $\eta_j = 1$, we have

$$\int_{\mathbb{R}^n} \frac{|\widehat{g}(x)|}{|x_1|\cdots|x_n|} \, dx = \int_{\mathbb{R}} \frac{1}{|x_1|} \left[\int_{\mathbb{R}^{n-1}} \frac{|\widehat{g}(x)|}{|x_2|\cdots|x_n|} \, dx_2 \cdots dx_n \right] dx_1$$

$$\lesssim \int_{\mathbb{R}} \frac{1}{|x_1|} \sum_{\substack{0\leq\eta\leq 1,\\ \eta_1=0}} \|\mathcal{H}_\eta F_{e_1} g(x_1,\cdot,\ldots,\cdot)\|_{L^1(\mathbb{R}^{n-1})} \, dx_1$$

$$= \sum_{\substack{0\leq\eta\leq 1,\\ \eta_1=0}} \int_{\mathbb{R}^{n-1}} \int_{\mathbb{R}} \frac{|F_{e_1}\mathcal{H}_\eta g(x_1,x_2,\ldots,x_n)|}{|x_1|} \, dx_1 \, dx_2 \cdots dx_n.$$

Applying now (1.61) with respect to x_1, we complete the proof. $\square$

Note that

$$H^1(\mathbb{R}\times\cdots\times\mathbb{R}) \subset H^1(\mathbb{R}^n), \quad (5.27)$$

see, e.g., [71, Th.1]. In the same paper [71], one can find A. Uchiyama's example that this inclusion is proper.

The perceptive reader could pay attention to the fact that unlike Chapter 1, where a one-dimensional background is given, in Chapter 5 which has a multidimensional background, as well as in what follows, no word on atomic or molecular characterization is said. The point is that the Hardy space we deal with there is of a product (mixed, multi-parameter, hybrid) nature, for which such a characterization faces serious difficulties. In a recent paper [78], a new atomic decomposition is suggested for such spaces. This may also open new horizons in regard to the problems in question.

Let us now introduce a multivariate version of Q which corresponds to our setting:

$$Q_H = \{g \in L^1(\mathbb{R}^n) : \|g\|_{Q_H} = \|g\|_{L^1(\mathbb{R}^n)} + \int_{\mathbb{R}^n} \frac{|\widehat{g}(x)|}{|x_1 \cdots x_n|}\, dx < \infty\}. \qquad (5.28)$$

As in dimension one, it will characterize widest spaces of integrability of the Fourier transforms for functions of bounded Hardy's variation.

5.5 Absolute continuity

In order to present multidimensional versions of Theorems 2.8 and 3.2, as well as extensions of others, we should discuss a multidimensional notion of absolute continuity; see, e.g., [22] or [67]. There are several definitions, the one that fits the functions of bounded Hardy variation will be used. Absolutely continuous functions will be defined as those which are representable as

$$f(x) = \int_{-\infty}^{x_1} \cdots \int_{-\infty}^{x_n} h(u)\, du + \sum_{\eta \neq \mathbf{1}} f_\eta(x_\eta), \qquad (5.29)$$

where marginal functions f_η depending on a smaller number of variables than n, in fact, $|\eta| < n$, since $|\eta| = n$ only if $\eta = \mathbf{1}$, are absolutely continuous on $\mathbb{R}^{|\eta|}$. More precisely, each f_η can be expressed by a formula similar to (5.29), with some function h_η in the corresponding integral

$$\left(\prod_{j:\eta_j=1}\right) \int_{-\infty}^{x_j} h_\eta(u_\eta)\, du_\eta.$$

In these terms it is possible, for consistency, to replace $h(u)$ in (5.29) with $h_{\mathbf{1}}(u)$.

This inductive definition is correct since it reduces to the usual absolute continuity on $\mathbb{R}$ for marginal functions of one variable. Local absolute continuity means absolute continuity on every finite rectangle $[a, b] = [a_1, b_1] \times \cdots \times [a_n, b_n]$. In this case, $a_1, \ldots, a_n$, respectively, should replace $-\infty$ in (5.29). In [180], a less restrictive version of absolute continuity is given in a similar way, without assuming the absolute continuity of the marginal functions in (5.29). However, this is not applicable in our considerations.

Like Proposition 1.18 in Chapter 1, the next statement "justifies" the study of the Fourier transform of a function with bounded Hardy's variation.

Proposition 5.30. *Let the integral*

$$\int_{\mathbb{R}^n} \frac{|\widehat{g}(x)|}{|x_1| \cdots |x_n|}\, dx$$

5.5. Absolute continuity

(the one on the left-hand side of (5.26)) be finite, or, equivalently, let $g \in Q_H$. Then g is the D^1 derivative of a function of bounded Hardy variation f which is absolutely continuous, $\lim_{|x| \to \infty} f(x) = 0$, and its Fourier transform is integrable.

Proof. Let
$$f(x) = \int_{-\infty}^{x_1} \cdots \int_{-\infty}^{x_n} g(u)\, du. \tag{5.31}$$

First of all, this immediately gives absolute continuity. For simplicity, let us continue the proof in dimension two. We have

$$\int_{-\infty}^{\infty} \int_{-\infty}^{\infty} g(s,t) e^{-ixs} e^{-iyt}\, ds\, dt$$

$$= \int_{-\infty}^{\infty} \left[e^{-iyt} \int_{-\infty}^{t} g(s,v)\, dv \right]_{-\infty}^{\infty} e^{-ixs}\, ds$$

$$+ iy \int_{-\infty}^{\infty} \left[\int_{-\infty}^{\infty} \int_{-\infty}^{t} g(s,v)\, dv e^{-iyt}\, dt \right] e^{-ixs}\, ds$$

$$= iy \int_{-\infty}^{\infty} \left[\int_{-\infty}^{\infty} \int_{-\infty}^{t} g(s,v)\, dv e^{-iyt}\, dt \right] e^{-ixs}\, ds.$$

Here the integrated term at infinity and similar ones in the sequel vanish because of the assumption of the proposition on the finiteness of the integral on the left-hand side of (5.26). We then repeat the same procedure with respect to the other variable. Noting that we are always able to change the order of operations because of the integrability of g, we get

$$iy \int_{-\infty}^{\infty} \left[\int_{-\infty}^{\infty} \int_{-\infty}^{t} g(s,v)\, dv e^{-iyt}\, dt \right] e^{-ixs}\, ds$$

$$= iy \int_{-\infty}^{\infty} \left[\int_{-\infty}^{\infty} \int_{-\infty}^{t} g(s,v)\, dv e^{-ixs}\, ds \right] e^{-iyt}\, dt$$

$$= iy \int_{-\infty}^{\infty} \left[e^{-ixs} \int_{-\infty}^{s} \int_{-\infty}^{t} g(u,v)\, dv\, du \right]_{-\infty}^{\infty} e^{-iyt}\, dt$$

$$+ ix\, iy \int_{-\infty}^{\infty} \left[\int_{-\infty}^{\infty} \int_{-\infty}^{s} \int_{-\infty}^{t} g(u,v)\, du\, dv e^{-ixs}\, ds \right] e^{-iyt}\, dt$$

$$= ix\, iy \int_{-\infty}^{\infty} \int_{-\infty}^{\infty} \left[\int_{-\infty}^{s} \int_{-\infty}^{t} g(u,v)\, du\, dv \right] e^{-ixs} e^{-iyt}\, ds\, dt$$

$$= ix\, iy \int_{-\infty}^{\infty} \int_{-\infty}^{\infty} f(s,t) e^{-ixs} e^{-iyt}\, ds\, dt.$$

It follows now from $g \in Q_H$ that the L^1 norm of $\widehat{f}$ coincides with the Q_H norm of the derivative of f. □

5.6 Integration by parts

We give a general integration by parts formula for the multiple Lebesgue-Stieltjes integral. For the case $n = 1$, see (1.8); more precisely,

$$\int_{a+}^{a+b} f(x) \, dg(x) = (fg)(a+b) - (fg)(a) - \int_{a+}^{a+b} g(x-) \, df(x);$$

the case $n = 2$ is outlined in [12]. The general formula below apparently appeared for the first time in [130] and proved to be useful in various calculations.

In the formulation and proof, we will denote

$$d_u f(x(\cdot); u(\cdot)) \quad \text{instead of} \quad d_{u(\cdot)} f(x(\cdot); u(\cdot)),$$

and so on. It will always be clear with respect to what we differentiate without unnecessary details. If we want to make clear over which coordinates we integrate in a multiple integral we write, e.g., for the $|\mathbb{N}_\chi|$-dimensional integral

$$\prod_{j:\eta_j \neq 0} \int_{a_j+}^{a_j+b_j} g(x_\chi; u_\eta-) \, d_u f(x_\chi; u_\eta),$$

slightly abusing the $\prod$-sign. Further, $a+ = (a_1+, \ldots, a_n+)$, while $u-$ similarly means approaching the point u from the left in each variable.

Lemma 5.32. *Assume that our functions are of Hardy bounded variation and are, without loss of generality, right continuous in each variable on $[a, a+b] \subset \mathbb{R}^n$. Then for Lebesgue–Stieltjes integrals, we have the following partial integration formula*

$$\int_{(a, a+b]} f(x) \, dg(x)$$

$$= \sum_{k=0}^{n} (-1)^{n-k} \sum_{\substack{\chi+\eta=1 \\ |\mathbb{N}_\chi|=\mathbf{k}}} \prod_{i:\chi_i \neq 0} -\Delta_{b_i} \prod_{j:\eta_j \neq 0} \int_{a_j+}^{a_j+b_j} g(a_\chi; x_\eta-) \, d_x f(a_\chi; x_\eta),$$

where $\Delta_{b_i} g(x) = g(x) - g(x+b_i e_i)$, and the term for $k = 0$ is just the n-dimensional integral, while $k = n$ is meant as

$$\prod_{k=1}^{n} -\Delta_{b_k} g(a) f(a).$$

Proof. The proof is based on induction; the case $n = 1$ is as described. The step from n to $n+1$ is as follows. Assume, without loss of generality, that $f(x) = 0$ if $x_i = a_i$ for some i. By Fubini's theorem, with $x, u, a, b \in \mathbb{R}^{n+1}$, we obtain

$$\int_{(a, a+b]} f(x) \, dg(x) = \int_{(a, a+b]} \int_{(a, x]} d_u f(u) \, d_v g(v)$$

$$= \left(\prod_{k=1}^{n} \int_{a_k+}^{a_k+b_k} \right) \left(\prod_{k=1}^{n} \int_{u_k+}^{a_k+b_k} \right) \int_{a_{n+1}+}^{a_{n+1}+b_{n+1}} \int_{u_{n+1}+}^{a_{n+1}+b_{n+1}} d_v g(v) \, d_u f(u).$$

5.6. Integration by parts

With $\widetilde{g}(\cdot) = g(\cdot; a_{n+1} + b_{n+1}) - g(\cdot; u_{n+1}-)$, we get by induction

$$\int_{(a,a+b]} f(x)\, dg(x)$$

$$= \int_{a_{n+1}+}^{a_{n+1}+b_{n+1}} \sum_{k=0}^{n} (-1)^{n-k} \sum_{\substack{\chi+\eta=\mathbf{1} \\ |\mathbb{N}_\chi|=k}} \prod_{i:\chi_i \neq 0} -\Delta_{b_i}$$

$$\times \prod_{j:\eta_j \neq 0} \int_{a_j+}^{a_j+b_j} \widetilde{g}(a_\chi; u_\eta-)\, d_u f(a_\chi; u_\eta; u_{n+1})$$

$$= \sum_{k=0}^{n} (-1)^{n-k} \sum_{\substack{\chi+\eta=\mathbf{1} \\ |\mathbb{N}_\chi|=k}} \prod_{i:\chi_i \neq 0} -\Delta_{b_i} \left[\prod_{j:\eta_j \neq 0} \int_{a_j+}^{a_j+b_j} \right.$$

$$\left(g(a_\chi; u_\eta-; a_{n+1} + b_{n+1})\, d_u f(a_\chi; u_\eta; a_{n+1} + b_{n+1}) \right.$$

$$- g(a_\chi; u_\eta-; a_{n+1} + b_{n+1})\, d_u f(a_\chi; u_\eta; a_{n+1})$$

$$- g(a_\chi; u_\eta-; a_{n+1})\, d_u f(a_\chi; u_\eta; a_{n+1} + b_{n+1})$$

$$\left. + g(a_\chi; u_\eta-; a_{n+1})\, d_u f(a_\chi; u_\eta; a_{n+1}) \right)$$

$$\left. - \int_{a_{n+1}+}^{a_{n+1}+b_{n+1}} \prod_{j:\eta_j \neq 0} \int_{a_j+}^{a_j+b_j} g(a_\chi; u_\eta-; u_{n+1})\, d_u f(a_\chi; u_\eta; u_{n+1}) \right]$$

$$= \sum_{k=0}^{n+1} (-1)^{n+1-k} \sum_{\substack{\chi+\eta=\mathbf{1} \\ |\mathbb{N}_\chi|=k}} \prod_{i:\chi_i \neq 0} -\Delta_{b_i}$$

$$\times \prod_{j:\eta_j \neq 0} \int_{a_j+}^{a_j+b_j} g(a_\chi; u_\eta-)\, d_u f(a_\chi; u_\eta),$$

where in the last two lines the vectors χ, η, and $\mathbf{1}$ are $n+1$-dimensional. This completes the proof. □

Chapter 6
Integrability of the Fourier transforms

In this chapter, we first obtain certain multidimensional results for the integrability of the Fourier transform (see (5.6))

$$\widehat{f}(x) = \int_{\mathbb{R}^n} f(u)e^{-i\langle x,u\rangle}du,$$

which extend some theorems from Part I. Let us mention that various results of that kind can be found in the survey paper [128]. For example, in Section 7 of that paper many results are extensions of Zygmund's test for the absolute convergence of Fourier series of a function of bounded variation. Concerning the results we are going to present here, our objective is not to generalize "everything". On the contrary, we restrict ourselves to a very modest task: to generalized very few results, mostly to demonstrate how the indicator notation works and to give some background for discretization results in Chapter 8. Moreover, one of the goals is to show how "bad" is the direct generalization of some results as compared with those in the next Chapter 7. The latter extends the advanced results from Chapter 4.

We also insert into this chapter extensions of the one-dimensional result of Section 2.6 in Chapter 2 concerning absolute continuity and the corresponding Hardy–Littlewood theorem.

6.1 Functions with derivatives in the Hardy type spaces

The first step obviously is to begin with a simple generalization of Theorem 3.2.

Theorem 6.1. *Let f be of bounded Hardy variation and let f vanish at infinity along with all $D^\eta f$ except for $\eta = 1$. Let also f and the same $D^\eta f$ be absolutely continuous. Then $\widehat{f} \in L^1(\mathbb{R}^n)$ if and only if $D^1 f \in Q_H$.*

Proof. In (5.6), we simply integrate by parts in each of the variables. Observe that we do need absolute continuity with respect to a part of variables here. The integrated terms vanish each time due to the assumptions of the theorem. After n steps we integrate
$$\frac{|\widehat{D^1 f}(x)|}{|x_1 \cdots x_n|}$$
in place of $\widehat{f}(x)$, which completes the proof. □

Corollary 6.2. *Let f be of bounded Hardy variation and let f vanish at infinity along with all $D^\eta f$ except for $\eta = \mathbf{1}$. Let also f and the same $D^\eta f$ be absolutely continuous. If $D^1 f \in H^1(\mathbb{R} \times \cdots \times \mathbb{R})$, then $\widehat{f} \in L^1(\mathbb{R}^n)$.*

Proof. We just apply the Fourier-Hardy inequality (5.26) to the integral that assigns, by virtue of the above theorem, that $D^1 f$ belongs to Q_H. □

Such an extension, a direct consequence of the Fourier-Hardy inequality, is not that developed as Theorem 2.8 for the sine case. Also, as was mentioned in the introduction, this is one of the cases where it is not necessary to assume that f is of bounded Hardy variation. Indeed, it is absolutely continuous and integrability of the corresponding derivative is guaranteed by the fact that $D^1 f \in H^1(\mathbb{R} \times \cdots \times \mathbb{R})$. We now give an analog of Theorem 2.8. It is not completely new; one can find similar versions in [107] (the same one is given in [92, Ch.3]). We hope that this version is more transparent, first of all because of indicator notation. Of course, this is the main reason why this chapter appears at all but there is an additional more mundane reason, to show how restrictive previous approaches are and why a new more precise one, is very much desirable. Also, just for this result, an analog for trigonometric series will be obtained in Chapter 8.

We consider
$$\widehat{f_\eta}(x) = \int_{\mathbb{R}^n_+} f(u) \prod_{i:\eta_i=1} \cos x_i u_i \prod_{i:\eta_i=0} \sin x_i u_i \, du. \tag{6.3}$$

If $\eta = \mathbf{1}$ we have the purely cosine transform, while if $\eta = \mathbf{0}$ we have the purely sine transform, otherwise we have a mixed transform with both cosines and sines.

Theorem 6.4. *Let $f : \mathbb{R}^n_+ \to \mathbb{C}$ be of bounded Hardy variation on $\mathbb{R}^n_+$ and let f vanish at infinity along with all $D^\eta f$ except $\eta = \mathbf{1}$. Let also f be locally absolutely continuous on $\mathbb{R}^n_+ \setminus \{0\}$. If $D^1 f \in H^1_0(\mathbb{R} \times \cdots \times \mathbb{R})$, then for $\eta = \mathbf{1}$ we have*
$$\int_{\mathbb{R}^n_+} |\widehat{f_1}(x)| \, dx \lesssim \|D^1 f\|_{H^1_0(\mathbb{R} \times \cdots \times \mathbb{R})}. \tag{6.5}$$

For $\eta \neq \mathbf{1}, \mathbf{0}$,
$$\int_{\mathbb{R}^n_+} |\widehat{f_\eta}(x)| \, dx \lesssim \sum_{\substack{\eta \neq 1,0 \\ 0 \leq \chi \leq \eta}} \sum_{0 \leq \chi \leq 1-\eta} \int_{\mathbb{R}^n_+} \frac{|\mathcal{H}_o^{\chi+\zeta} D^{\eta+\chi} f(x)|}{\prod_{j:\chi_j=\eta_j=0} x_j} \, dx; \tag{6.6}$$

6.1. Functions with derivatives in the Hardy type spaces

and for $\eta = \mathbf{0}$,

$$\widehat{f}_{\mathbf{0}}(x) = f\left(\frac{\pi}{2x_1}, \ldots, \frac{\pi}{2x_n}\right) \prod_{j=1}^{n} \frac{1}{x_j} + F(x), \tag{6.7}$$

with

$$\int_{\mathbb{R}_+^n} |F(x)|\, dx \lesssim \sum_{\substack{0 \leq \chi \leq 1, \chi \neq 0 \\ 0 \leq \varsigma \leq \chi}} \int_{\mathbb{R}_+^n} \frac{|\mathcal{H}_o^{\varsigma} D^{\chi} f(x)|}{\prod_{j: \chi_j = 0} x_j}\, dx. \tag{6.8}$$

Proof. We start with the proof in dimension two. The only cosine case is simple for any dimension: we integrate by parts n times, in each of the variables, and then apply (5.26).

The following two cases in dimension two illustrate along which lines the proof runs in other situations. In the mixed case, integrating by parts, we obtain

$$\widehat{f}_{(1,0)}(x, y) = \int_{\mathbb{R}_+^2} f(u, v) \cos xu \sin yv\, du\, dv$$

$$= -\frac{1}{x} \int_{\mathbb{R}_+^2} D^{(1,0)} f(u, v) \sin xu \sin yv\, du\, dv.$$

Applying now (1.61) in x and then (2.9), the cosine part of Theorem 2.8, we have

$$\int_0^\infty |\widehat{f}_{(1,0)}(x, y)|\, dx \lesssim \int_0^\infty \left|\int_0^\infty D^{(1,0)} f(x, v) \sin yv\, dv\right| dx$$

$$+ \int_0^\infty \left|\int_0^\infty \mathcal{H}_o^{(1,0)} D^{(1,0)} f(x, v) \sin yv\, dv\right| dx.$$

Applying the sine part of Theorem 2.8 to the inner integrals on the right-hand side, we finally get

$$\int_0^\infty \int_0^\infty |\widehat{f}_{(1,0)}(x, y)|\, dx\, dy \lesssim \int_{\mathbb{R}_+^2} \frac{|D^{(1,0)} f(x, y)|}{y}\, dx\, dy$$

$$+ \int_{\mathbb{R}_+^2} |D^{(1,1)} f(x, y)|\, dx\, dy$$

$$+ \int_{\mathbb{R}_+^2} \frac{|\mathcal{H}_o^{(1,0)} D^{(1,0)} f(x, y)|}{y}\, dx\, dy$$

$$+ \int_{\mathbb{R}_+^2} |\mathcal{H}_o^{(0,1)} D^{(1,1)} f(x, y)|\, dx\, dy$$

$$+ \int_{\mathbb{R}_+^2} |\mathcal{H}_o^{(1,0)} D^{(1,1)} f(x, y)|\, dx\, dy$$

$$+ \int_{\mathbb{R}_+^2} |\mathcal{H}_o^{(1,1)} D^{(1,1)} f(x, y)|\, dx\, dy.$$

Let us proceed to the "only sine" case. Again, by the sine part of Theorem 2.8, we get

$$\widehat{f}_{(0,0)}(x) = \int_{\mathbb{R}_+^2} f(u,v) \sin xu \sin yv \, du \, dv$$

$$= \frac{1}{y} \int_0^\infty f\left(u, \frac{\pi}{2y}\right) \sin xu \, du + F_2(x,y), \tag{6.9}$$

with

$$\int_0^\infty |F_2(x,y)| \, dy \lesssim \int_0^\infty \left| \int_0^\infty D^{(0,1)} f(u,y) \sin xu \, du \right| dy$$

$$+ \int_0^\infty \left| \mathcal{H}_{(0,1)}^o \int_0^\infty D^{(0,1)} f(u,y) \sin xu \, du \right| dy.$$

Applying one more time the sine part of Theorem 2.8 to the leading term on the right-hand side of (6.9) and to the inner integrals that give estimates for F_2, we finally obtain

$$\widehat{f}_{(0,0)}(x,y) = \int_{\mathbb{R}_+^2} f(u,v) \sin xu \sin yv \, du \, dv$$

$$= \frac{1}{x}\frac{1}{y} f\left(\frac{\pi}{2x}, \frac{\pi}{2y}\right) + F(x,y), \tag{6.10}$$

with

$$\int_0^\infty \int_0^\infty |F(x,y)| \, dx \, dy$$

$$\lesssim \int_{\mathbb{R}_+^2} \frac{|D^{(1,0)} f(x,y)|}{y} \, dx \, dy + \int_{\mathbb{R}_+^2} |D^{(1,1)} f(x,y)| \, dx \, dy$$

$$+ \int_{\mathbb{R}_+^2} \frac{|D^{(0,1)} f(x,y)|}{x} \, dx \, dy + \int_{\mathbb{R}_+^2} \frac{|\mathcal{H}_o^{(1,0)} D^{(1,0)} f(x,y)|}{y} \, dx \, dy$$

$$+ \int_{\mathbb{R}_+^2} \frac{|\mathcal{H}_o^{(0,1)} D^{(0,1)} f(x,y)|}{y} \, dx \, dy + \int_{\mathbb{R}_+^2} |\mathcal{H}_o^{(0,1)} D^{(1,1)} f(x,y)| \, dx \, dy$$

$$+ \int_{\mathbb{R}_+^2} |\mathcal{H}_o^{(1,0)} D^{(1,1)} f(x,y)| \, dx \, dy + \int_{\mathbb{R}_+^2} |\mathcal{H}_o^{(1,1)} D^{(1,1)} f(x,y)| \, dx \, dy.$$

Now, only some additional explanation about the general case is in order. If $\eta \neq \mathbf{1}, \mathbf{0}$, then we integrate by parts $|\eta|$ times, in each of the variables u_j where $\eta_j = 1$ and apply (5.26). For the rest of the variables, we apply, step by step, the sine part of Theorem 2.8, as above in the two-dimensional purely sine case. The latter works in the purely sine case, that is, if $\eta = \mathbf{0}$, as well, but in

6.1. Functions with derivatives in the Hardy type spaces

addition to the estimates we obtain the leading term in (6.7). In the estimates (6.8) as well as in the "sine" part of (6.5), each of the variables is subject to one of the following three operations: either division by the argument with no differentiation, or differentiation, or the (odd) Hilbert transform of the derivative. The only difference from that given above with details the two-dimensional case is that two variables are not enough for all three operations together. They appear at most in pairs, while for the larger dimension all the combinations may appear simultaneously. These affect only notation in the sense that additional indicator vectors are involved. However, the proof itself goes along the same lines. $\square$

Applying (1.93) in each of the needed variables, we can establish the following consequences of the above theorems. Written in a somewhat different form (less convenient in our opinion) they can be found in [107], [128, Th.12.10] and in [92, Ch.3]; for those proved in dimension two in full detail and directly rather than via embeddings, see [68], [67], [69].

Corollary 6.11. *Let $f : \mathbb{R}_+^n \to \mathbb{C}$ be of bounded Hardy variation on $\mathbb{R}_+^n$ and let f vanish at infinity along with all $D^\eta f$ except $\eta = \mathbf{1}$. Let also f be locally absolutely continuous on $\mathbb{R}_+^n \setminus \{0\}$. Then for $\eta \neq \mathbf{0}$, we have*

$$\int_{\mathbb{R}_+^n} |\widehat{f}_\eta(x)|\, dx \lesssim \sum_{\mathbf{0} \leq \chi \leq \mathbf{1}-\eta} \int_{\mathbb{R}_+^n} \prod_{i:\chi_i=1} \frac{1}{x_i}$$
$$\left[\left(\prod_{j:\chi_j=0} \frac{1}{x_j} \int_{x_j \leq u_j \leq 2x_j} du_j \right) |D^{\mathbf{1}-\chi} f(u_{\mathbf{1}-\chi}, x_\chi)|^q \right]^{\frac{1}{q}} dx, \quad (6.12)$$

and for $\eta = \mathbf{0}$, we have (6.7), with

$$\int_{\mathbb{R}_+^n} |F(x)|\, dx \lesssim \int_{\mathbb{R}_+^n} \left[\left(\prod_{j=1}^n \frac{1}{x_j} \int_{x_j \leq u_j \leq 2x_j} \right) |D^{\mathbf{1}} f(u)|^q\, du \right]^{\frac{1}{q}} dx, \quad (6.13)$$

for some $1 < q < \infty$.

Corollary 6.14. *Let $f : \mathbb{R}_+^n \to \mathbb{C}$ be of bounded Hardy variation on $\mathbb{R}_+^n$ and let f vanish at infinity along with all $D^\eta f$ except $\eta = \mathbf{1}$. Let also f be locally absolutely continuous on $\mathbb{R}_+^n \setminus \{0\}$. Then for $\eta \neq \mathbf{0}$, we have*

$$\int_{\mathbb{R}_+^n} |\widehat{f}_\eta(x)|\, dx \lesssim \sum_{\mathbf{0} \leq \chi \leq \mathbf{1}-\eta} \int_{\mathbb{R}_+^n} \prod_{i:\chi_i=1} \frac{1}{x_i} \operatorname*{ess\,sup}_{\substack{x_j \leq u_j \leq 2x_j \\ j:\chi_j=0}} |D^{\mathbf{1}-\chi} f(u_{\mathbf{1}-\chi}, x_\chi)|\, dx, \quad (6.15)$$

and for $\eta = \mathbf{0}$, we have (6.7), with

$$\int_{\mathbb{R}_+^n} |F(x)|\, dx \lesssim \int_{\mathbb{R}_+^n} \operatorname*{ess\,sup}_{\substack{x_j \leq u_j \leq 2x_j \\ j=1,2,\ldots,n}} |D^{\mathbf{1}} f(u)|\, dx. \quad (6.16)$$

It should be mentioned that (6.12) and (6.13) as well as (8.34) and (8.35) are a mixture of two types of estimates. However, in calculations the third type appears only the L^1 norm with respect to certain variables. It suffices to demonstrate that the mixture of it and the O_q type norms is controlled by the O_q type norm only for two variables. We have

$$\int_0^\infty \int_0^\infty \left(\frac{1}{y}\int_y^{2y} |g(x,t)|^q\,dt\right)^{\frac{1}{q}} dy\,dx$$

$$= \frac{1}{\ln 2}\int_0^\infty \int_0^\infty \frac{1}{x}\int_x^{2x} \left(\frac{1}{y}\int_y^{2y} |g(s,t)|^q\,dt\right)^{\frac{1}{q}} ds\,dx\,dy.$$

Applying Hölder's inequality to the integral over $[x, 2x]$, we show that the right-hand side is dominated by

$$\int_0^\infty \int_0^\infty \left(\frac{1}{xy}\int_x^{2x}\int_y^{2y} |g(s,t)|^q\,ds dt\right)^{\frac{1}{q}} dx\,dy.$$

It is clear that Theorem 2.20 and its consequences can be generalized in the same manner. Since this chapter is somewhat "negative", by illustration of the disadvantages of the undertaken approach before demonstrating a more successful one in the next chapter, we do not need to dramatize the situation and thus omit such generalizations. What is even more striking is the situation with the possible extension of Theorem 3.5. Its form immediately tells us that a strait-forward extension might be powerless. All these lead us to proceed to multidimensional generalizations of theorems obtained in Chapter 4; in fact, they were meant for this. Their extensions is the content of the next chapter.

6.2 Absolute continuity, integrability of the Fourier transform and a Hardy–Littlewood theorem

In this section, we present extensions of Theorem 2.53 (see [134]). We start with the needed preliminary results, generalizations of Propositions 2.43 and 2.51, respectively, that are of interest in their own right.

6.2.1 Commutativity

One of the important ingredients in our study is the following generalization of Proposition 2.43. Of course, $\widetilde{\mathcal{H}}_\eta$ is understood exactly as the repeated ($|\eta|$ times) modified Hilbert transform (1.63).

Theorem 6.17. *If f and $\widetilde{\mathcal{H}}_\zeta f$ are $AC_{\eta+\chi}$ for all $0 < \zeta < \eta$, $0 < \eta \leq 1$ and χ such that $\eta + \chi \leq 1$, and $VH_{\eta+\chi}(\widetilde{\mathcal{H}}_\eta f) < \infty$, then*

$$D^{\eta+\chi}\widetilde{\mathcal{H}}_\eta f(x) = D^\eta \widetilde{\mathcal{H}}_\eta D^\chi f(x) = \mathcal{H}_\eta D^{\eta+\chi} f(x) \qquad (6.18)$$

6.2. Hardy–Littlewood theorem

for almost every $x \in \mathbb{R}^n$. In particular, if $\widetilde{\mathcal{H}}_\zeta f$ are AC for all $0 < \zeta < \eta$, $0 < \eta \leq 1$, and $VH(\widetilde{\mathcal{H}}_\eta f) < \infty$, then almost everywhere

$$D^1 \widetilde{\mathcal{H}}_\eta f(x) = \mathcal{H}_\eta D^1 f(x). \tag{6.19}$$

Proof. The first equality in (6.18) is simple. This is a question of changing the orders of two limits: the one that defines the principal value in the conjugate functions and the other that defines the D^χ derivative. The repeated limits exist in each order, since, on the one hand, we assume the "lower" conjugate functions to be absolutely continuous and the η-th of bounded Hardy's variation, and on the other hand, the conjugate functions of D^χ exist a.e., since D^χ is absolutely continuous with respect to the corresponding variables. That they may not coincide on a set of at most measure zero is proved exactly as in the one-dimensional case by means of Egorov's theorem.

With the first equality in (6.18) in hand, we prove the second one by induction. The one-dimensional base is immediately provided by Theorem 2.43. Indeed, for $\eta = e_j$, $j = 1, 2, \ldots, n$, we have $\chi_j = 0$ and $D^\chi f(x) \in AC_j$. Now, $\widetilde{\mathcal{H}}_\eta D^\chi f$ is such since it is equal to $D^\chi \widetilde{\mathcal{H}}_\eta f$ a.e., which is such by assumptions of the theorem. Assume now that the assertion is true for $\eta \neq 1$ and prove it for $\eta + e_j$, with j such that $\eta_j = \chi_j = 0$. We have

$$D^{\eta+e_j} \widetilde{\mathcal{H}}_{\eta+e_j} D^\chi f = D^j D^\eta \widetilde{\mathcal{H}}_j \widetilde{\mathcal{H}}_\eta D^\chi f = D^j \widetilde{\mathcal{H}}_j D^\eta \widetilde{\mathcal{H}}_\eta D^\chi f,$$

by the first equality in (6.18). Here η acts as χ in the theorem, and $D^\chi f, \widetilde{\mathcal{H}}_\eta D^\chi f \in AC_j$, by assumptions of the theorem. Similarly, the right-hand side is

$$\mathcal{H}_j D^j D^\eta \widetilde{\mathcal{H}}_\eta D^\chi f = \mathcal{H}_j D^\eta \widetilde{\mathcal{H}}_\eta D^{\chi+e_j} f.$$

Finally, by the assumption of induction, the latter is

$$\mathcal{H}_j \mathcal{H}_\eta D^\eta D^{\chi+e_j} f = \mathcal{H}_{\eta+e_j} D^{\eta+\chi+e_j} f,$$

as required. $\square$

6.2.2 Conditions for absolute continuity

Here we present a multidimensional analog of a theorem of F. and M. Riesz on absolute continuity, more precisely, an extension of Theorem 2.51.

Theorem 6.20. *Let $VH(f) < \infty$ and $VH(\widetilde{\mathcal{H}}_\eta f) < \infty$ for all $0 < \eta \leq 1$. Then f and $\widetilde{\mathcal{H}}_\eta f$ all are AC.*

Proof. We use inductive arguments. Theorem 2.51 forms the basis for every $\eta = e_j$, $j = 1, \ldots, n$. Assuming that the result is true for some n, we prove it for $n + 1$. Let $x = (x_1, \ldots, x_n)$ and $(x, x_{n+1}) \in \mathbb{R}^{n+1}$. Let the indicator vectors η and $\mathbf{1}$ continue to be n-dimensional, that is, with respect to x. If $VH(f) < \infty$ in $\mathbb{R}^{n+1}$, then f has bounded Hardy variation also with respect to x. The same are all $\widetilde{\mathcal{H}}_\eta f$. Therefore, by the induction assumption, f and all $\widetilde{\mathcal{H}}_\eta f$ are absolutely continuous with respect to x.

Now, $VH(f) < \infty$ in $\mathbb{R}^{n+1}$ yields that $D^1 f$ is of bounded variation with respect to x_{n+1}. Similarly, $VH(\widetilde{\mathcal{H}}_{n+1}f) < \infty$ in $\mathbb{R}^{n+1}$ and thus $\widetilde{\mathcal{H}}_{n+1}D^1 f$ is of bounded variation with respect to x_{n+1}. By Theorem 2.51, $D^1 f$ is absolutely continuous with respect to x_{n+1}, and hence f is absolutely continuous in $\mathbb{R}^{n+1}$.

The absolute continuity for the conjugate functions is proved similarly. For example, we derive $D^1\widetilde{\mathcal{H}}_1 f$ to be of bounded variation with respect to x_{n+1} from $VH(\widetilde{\mathcal{H}}_1 f) < \infty$ in $\mathbb{R}^{n+1}$ and

$$D^1 \widetilde{\mathcal{H}}_{n+1} \widetilde{\mathcal{H}}_1 f = \widetilde{\mathcal{H}}_{n+1} D^1 \widetilde{\mathcal{H}}_1 f$$

to be of bounded variation with respect to x_{n+1} from $VH(\widetilde{\mathcal{H}}_{n+1}\widetilde{\mathcal{H}}_1 f) < \infty$ in $\mathbb{R}^{n+1}$. Again, from Theorem 2.51 we derive $D^1 \widetilde{\mathcal{H}}_1 f \in AC$ with respect to x_{n+1}, and with $\widetilde{\mathcal{H}}_1 f \in AC$ with respect to x in hand, we get $\widetilde{\mathcal{H}}_1 f \in AC$ in $\mathbb{R}^{n+1}$. □

This theorem asserts in particular that the previous theorem is meaningful, that is, assumptions of absolute continuity in it are natural. However, they can be proved in either order, independently.

6.2.3 Hardy–Littlewood type theorems

The following extension of Theorem 2.53 comes to mind.

Theorem 6.21. *Let f and $\widetilde{\mathcal{H}}_\eta f$, $0 < \eta \leq 1$, be of bounded Hardy variation. Let, in addition, VV_η of them vanish at infinity for all η except $\eta = 1$, that is, $\lim_{x_{1-\eta} \to \infty} VV_\eta(x_{1-\eta}) = 0$. Then the Fourier transforms of all of them are integrable.*

Indeed, with the results of the previous sections in hand, the proof goes along the same lines as that in dimension one.

Proof. Since the function f is of bounded Hardy variation, its derivatives $D^\eta f$ exist almost everywhere and each is integrable over $\mathbb{R}^\eta$. The assumptions of the theorem allow one to apply Theorem 6.20 with $\eta = 1$ and conclude that $f \in AC$. Using now Theorem 6.17, we see that for every η, $0 < \eta \leq 1$,

$$D^1 \widetilde{\mathcal{H}}_\eta f = \mathcal{H}_\eta D^1 f,$$

and the right-hand side is integrable over $\mathbb{R}^n$. Therefore $D^1 f \in H^1(\mathbb{R} \times \cdots \times \mathbb{R})$. Integrating by parts, which is possible since f is absolutely continuous, we obtain

$$\widehat{D^1 f}(x) = \int_{\mathbb{R}^n} D^1 f(u) e^{-i\langle x, u\rangle} du = \left(\prod_{j=1}^n ix_j\right) \int_{\mathbb{R}^n} f(u) e^{-i\langle x, u\rangle} du.$$

6.2. Hardy–Littlewood theorem

Dividing both sides by $\left(\prod_{j=1}^{n} ix_j\right)$, we derive from (5.26) that

$$\|\widehat{f}\|_{L^1(\mathbb{R}^n)} \lesssim \|D^1 f\|_{H^1(\mathbb{R}\times\cdots\times\mathbb{R})}.$$

Further, we have

$$\left(\prod_{j=1}^{n} i\operatorname{sign} x_j\right) \widehat{D^1 f}(x) = \widehat{\mathcal{H}_1 D^1 f}(x)$$

(see, e.g., [69]), which, by Theorem 6.17 with $\eta = 1$, is the Fourier transform of $D^1 \widetilde{\mathcal{H}}_1 f$. Integrating by parts as above, we conclude that in our situation the left-hand side of (5.26) is also the L^1 norm of the Fourier transform of $\widetilde{\mathcal{H}}_1 f$. Integrability of the rest of conjugate functions can be established along the same lines. The proof is complete. □

Chapter 7
Sharp results

Some results in the previous chapter, to be exact, the extensions of Theorem 2.8 and its consequences, have partially been presented in order to show how good they are but, at the same time, how restricted they are. More specifically, to show how much the remainder terms in the one-dimensional prototypes affect the multivariate relations. In this chapter the most precise multidimensional extensions are given. The needed machinery for this has been described in Chapter 4. We shall deal with

$$\widehat{f}_\eta(x) = \int_{\mathbb{R}_+^n} f(u) \prod_{i:\eta_i=1} \cos x_i u_i \prod_{i:\eta_i=0} \sin x_i u_i \, du,$$

as in (6.3), where f is a locally absolutely continuous function with bounded Hardy variation. The question, why among a variety of the notions of multidimensional variation the one due to Hardy (and Krause) is the most natural for the considered problems, has been discussed in Chapter 5.

Recall that if $\eta = \mathbf{1}$ in (6.3) we have the pure cosine transform, while if $\eta = \mathbf{0}$ we have the pure sine transform, otherwise we have a mixed transform with both cosines and sines. But even in the case where the Fourier transform is the sine in each variable, the only case where a sort of multidimensional asymptotic formula has been known till recently (see previous chapter), the remainder terms are rough in a sense. In order to get more advanced multivariate generalizations, the asymptotic relations in Theorems 4.15, 4.20 and 4.23 are obtained. They have much more delicate form than those in Theorems 2.8, 2.20 and 3.5. For this, an operator balancing all the terms in these asymptotic relations is introduced, see (1.122). Recall that it is defined by means of a generating function φ and takes the value

$$B_\varphi g(x) = \frac{1}{x^2} \int_0^\infty g\left(\frac{t}{x}\right) \varphi(t) \, dt$$

on an appropriate function g. Its only (obvious) property that we will need is that for $g \in L^1(\mathbb{R}_+)$, we have $B_\varphi \in L^1(\mathbb{R}_+)$ provided

$$\int_0^\infty \frac{|\varphi(t)|}{t}\, dt < \infty,$$

which is (1.127) in Corollary 1.126. Recall that one more peculiarity that makes new one-dimensional and multi-dimensional results different from the preceding ones is that, in fact, we do not assume that f' in dimension one and corresponding derivatives in several dimensions belong to some Hardy space. This can be done in applications, say, in order to simplify calculations but general relations do not claim for this. What is assumed instead is a much less restrictive assumption that the inverse formula for the Hilbert transform (see either (4.1) or (4.2)) of f' holds true almost everywhere. In the multivariate case, we assume this with respect to appropriate variables by just writing that the inverse formula for the Hilbert transform holds, without indicating a specific relation.

In general, the results we are going to obtain in this chapter will be of the form

$$\widehat{f}_\eta(x) = \textbf{many leading terms} + \text{integrable remainder terms}.$$

Some of the leading terms as well as the remainder terms will be combinations of balance operators generated by various special kernels φ.

As in the corresponding situation with one-dimensional results, we begin with reconsidering convexity type results. They are very recent, see [127]. As in dimension one, a multivariate generalization of the Sz.-Nagy type theorem (see [120]) is well-suited for this chapter.

7.1 Convexity type results

Surprisingly, though many of the results on the Fourier transform of a function of bounded variation were extended to several dimensions (see, e.g., [107], [122], [124], [125]), the basic results with convexity were generalized only in very particular cases, say, for radial functions (see, e.g., [92, Chapter 4] or [121]).

In the multidimensional case, the balance operator will be used in full generality.

We study, for $\gamma = (\gamma_1, \ldots, \gamma_n)$, $\gamma_j = 0$ or 1, $j = 1, 2, \ldots, n$, the Fourier transforms

$$\widehat{f}_\gamma(x) = \int_{\mathbb{R}_+^n} f(u) \left(\prod_{j=1}^n \cos\left(x_j u_j - \frac{\pi \gamma_j}{2}\right) \right) du. \qquad (7.1)$$

It is clear that $\widehat{f}_\gamma$ represents the cosine Fourier transforms in the variables where $\gamma_j = 0$, while taking $\gamma_j = 1$ gives the sine component.

An analog of Theorem 4.5 reads as follows.

7.1. Convexity type results

Theorem 7.2. *Let f be locally absolutely continuous on $(a,b) \subset \mathbb{R}_+^n$ and let*

$$d_j = \min\left\{\frac{b_j - a_j}{2}, \frac{\pi}{2}\right\}, \quad j = 1, 2, \ldots, n.$$

Then for $x_j > 0$, $j = 1, 2, \ldots, n$,

$$\widehat{f}_\gamma(x) = \left(\prod_{j=1}^n \frac{1}{x_j}\right) \sum_{0 \le \eta \le 1} (-1)^{|\eta|} \left\{\prod_{j:\eta_j=1} \sin(b_j x_j - \frac{\pi\gamma_j}{2})\right.$$

$$\left. \times \prod_{j:\eta_j=0} \sin(a_j x_j - \frac{\pi\gamma_j}{2})\right\} f\left((b - \frac{d}{x})_\eta, (a + \frac{d}{x})_{1-\eta}\right)$$

$$+ \sum_{0 < \chi < 1} \left(\prod_{j:\chi_j=1} \frac{1}{x_j}\right) \sum_{\substack{\varsigma:\varsigma_i = 0 \text{ if } \chi_i = 0 \\ 0 \le \varsigma \le \chi}} (-1)^{|\varsigma|}$$

$$\times \left\{\prod_{j:\varsigma_j=1} \sin(b_j x_j - \frac{\pi\gamma_j}{2}) \prod_{j:\eta_j=0, \chi_j=1} \sin(a_j x_j - \frac{\pi\gamma_j}{2})\right\}$$

$$\times B_K^{1-\chi} D^{1-\chi} f\left((b - \frac{d}{x})_\varsigma, (a + \frac{d}{x})_{\chi-\varsigma}, x_{1-\chi}\right)$$

$$+ B_K^1 D^1 f(x). \tag{7.3}$$

Proof. With all the preliminaries, including the one-dimensional result, the proof is simple. Assume that $f(u) = f_1(u_1) \cdots f_n(u_n)$, where each f_j satisfies the assumptions of Theorem 4.5. Then the answer will be just the product of (4.6) applied to each f_j. But the general result, due to the assumptions, is the same! The only difference is that instead of the product of functions, we have the product (composition) of operators B_K and D^η for certain summands. Technically, we apply (4.6) in one of the variables, getting thereby three summands, each depending on the rest of the variables. Then we just repeat this procedure $n - 1$ times. □

In dimension two, for instance, (7.3) looks as follows.

$$\int_{a_1}^{b_1}\int_{a_2}^{b_2} f(u_1, u_2) \cos(x_1 u_1 - \frac{\pi\gamma_1}{2}) \cos(x_2 u_2 - \frac{\pi\gamma_2}{2}) \, du_1 \, du_2$$

$$= \frac{1}{x_1 x_2}\left\{f\left(b_1 - \frac{d_1}{x_1}, b_2 - \frac{d_2}{x_2}\right) \sin(b_1 x_1 - \frac{\pi\gamma_1}{2}) \sin(b_2 x_2 - \frac{\pi\gamma_2}{2})\right.$$

$$- f\left(b_1 - \frac{d_1}{x_1}, a_2 + \frac{d_2}{x_2}\right) \sin(b_1 x_1 - \frac{\pi\gamma_1}{2}) \sin(a_2 x_2 - \frac{\pi\gamma_2}{2})$$

$$- f\left(a_1 + \frac{d_1}{x_1}, b_2 - \frac{d_2}{x_2}\right) \sin(a_1 x_1 - \frac{\pi\gamma_1}{2}) \sin(b_2 x_2 - \frac{\pi\gamma_2}{2})$$

$$+ f\left(a_1 + \frac{d_1}{x_1}, a_2 + \frac{d_2}{x_2}\right)\sin(a_1 x_1 - \frac{\pi\gamma_1}{2})\sin(a_2 x_2 - \frac{\pi\gamma_2}{2})\Big\}$$

$$+ \frac{1}{x_2}\sin(b_2 x_2 - \frac{\pi\gamma_2}{2}) B_K^{(1,0)} D^{(1,0)} f(x_1, b_2 - \frac{d_2}{x_2})$$

$$- \frac{1}{x_2}\sin(a_2 x_2 - \frac{\pi\gamma_2}{2}) B_K^{(1,0)} D^{(1,0)} f(x_1, a_2 + \frac{d_2}{x_2})$$

$$- \frac{1}{x_1}\sin(a_1 x_1 - \frac{\pi\gamma_1}{2}) B_K^{(0,1)} D^{(0,1)} f(a_1 + \frac{d_1}{x_1}, x_2)$$

$$+ \frac{1}{x_1}\sin(b_1 x_1 - \frac{\pi\gamma_1}{2}) B_K^{(0,1)} D^{(0,1)} f(b_1 - \frac{d_1}{x_1}, x_2)$$

$$+ B_K^{(1,1)} D^{(1,1)} f(x_1, x_2).$$

7.1.1 Functions of convex type

As in dimension one, we show that some terms in (7.3) can be made integrable under certain convexity type assumptions. It is more or less predictable that the usual multidimensional convexity is not enough. We introduce the notion of robust convexity, again assuming both convexity downwards and convexity upwards. The classical definition of a convex function f of several variables defined on a convex subset $G \subset \mathbb{R}^n$ means that

$$f((1-\lambda)x_0 + \lambda x_1) \leq (1-\lambda)f(x_0) + \lambda f(x_1)$$

for any $x_0, x_1 \in G$ and $\lambda \in [0,1]$ (see, e.g., [142, Appendix IV] or [154, 3.4] for this definition and properties). We may equally consider the opposite one with $\geq$ in place of $\leq$.

Definition 7.4. We say that a real-valued function f defined on $[a,b] \subset \mathbb{R}^n$ is *robust convex* if it is convex on $[a,b]$ in each variable with the rest of them fixed, and for any η, $\mathbf{0} < \eta \leq \mathbf{1} - \mathbf{e_j}$, $j = 1, 2, \ldots, n$, the function $D^\eta f(u)$ is convex in each of the variables of $u_{1-\eta}$ with the rest of $n-1$ variables fixed.

Regarding the first part of this definition, we mention that the function convex in the above (classical) sense is convex in every direction but the function convex in each variable is not necessarily convex (see, e.g., [154, 3.4]).

In fact, only the first 2^n terms on the right-hand side of (7.3) (the first line in (7.3)) are always leading, in a sense. Similarly, only the last one there, $B_K^1 D^1 f(x)$, is a "genuine" remainder in many situations. Of course, the intermediate terms may play both roles, each of them is "less natural" than aforementioned ones. With these remarks in hand, we formulate and prove a very general extension of Theorem 4.3.

Theorem 7.5. *Let f be a continuous function defined on $[a,b]$, $0 \leq a_j < b_j \leq +\infty$, $j = 1, \ldots, n$, robust convex on $[a,b]$. Then for each x, with $x_j > 2$, $j = 1, \ldots, n$,*

7.1. Convexity type results

we have (7.3), with

$$\int_{\mathbb{R}^n_+ \setminus [0,2]} |B^1_K D^1 f(x)| \leq C \int_{[a,b]} |D^1 f(x)| \, dx.$$

Proof. As in Subsection 4.1.2, we estimate the corresponding integrals for the ingredients of K before the change of variables. To see how this works, two-dimensional procedures suffice completely. We shall consider in detail the most typical of them. To this end, we estimate

$$\int_2^\infty \int_2^\infty \frac{1}{x_1 x_2} \left| \int_{a_1}^{a_1 + \frac{d_1}{x_1}} \int_{a_2 + \frac{d_2}{x_2}}^{b_2 - \frac{d_2}{x_2}} D^{(1,1)} f(u_1, u_2) \left[\sin(a_1 x_1 - \frac{\pi \gamma_1}{2}) \right. \right.$$
$$\left. \left. - \sin(x_1 u_1 - \frac{\pi \gamma_1}{2}) \right] \sin(x_2 u_2 - \frac{\pi \gamma_2}{2}) \, du_1 du_2 \right| dx_1 dx_2$$

$$\leq \int_2^\infty \frac{1}{x_2} \int_2^\infty \int_{a_1}^{a_1 + \frac{d_1}{x_1}} (u_1 - a_1) \left| \int_{a_2 + \frac{d_2}{x_2}}^{b_2 - \frac{d_2}{x_2}} D^{(1,1)} f(u_1, u_2) \sin(x_2 u_2 \right.$$
$$\left. - \frac{\pi \gamma_2}{2}) \, du_2 \right| du_1 dx_1 \, dx_2.$$

Repeating (4.10) for the first variable, we get

$$d_1 \int_{a_1}^{a_1+d_1} \int_2^\infty \frac{1}{x_2} \left| \int_{a_2+\frac{d_2}{x_2}}^{b_2 - \frac{d_2}{x_2}} D^{(1,1)} f(u_1, u_2) \sin(x_2 u_2 - \frac{\pi \gamma_2}{2}) \, du_2 \right| dx_2 \, du_1.$$

Repeating now (4.11) and the calculations thereafter, we complete the estimate.
It is also worthwhile to consider the following case:

$$\int_2^\infty \int_2^\infty \frac{1}{x_1 x_2} \left| \int_{a_1 + \frac{d_1}{x_1}}^{b_1 - \frac{d_1}{x_1}} \int_{a_2 + \frac{d_2}{x_2}}^{b_2 - \frac{d_2}{x_2}} D^{(1,1)} f(u_1, u_2) \sin(x_1 u_1 - \frac{\pi \gamma_1}{2}) \right.$$
$$\left. \sin(x_2 u_2 - \frac{\pi \gamma_2}{2}) \, du_1 du_2 \right| dx_1 dx_2.$$

Applying (4.11) twice, in each of the variables, we arrive at

$$\int_{a_1 + \frac{d_1}{x_1}}^{b_1 - \frac{d_1}{x_1}} \int_{a_2 + \frac{d_2}{x_2}}^{b_2 - \frac{d_2}{x_2}} |d^{(1,1)} D^{(1,1)} f(u_1, u_2)|$$

in the inner integrals. But Definition 7.4 is perfectly tailored for transferring the absolute value operation outside the integrals, exactly as in (4.12) and (4.13). As is mentioned, the whole proof goes along these lines. □

Remark 7.6. If we try to formulate a less general but apparently more transparent assertion by leaving only the terms of the first line in (7.3) to be leading, the values like

$$\int_{[a,b]} \left(\prod_{j:\chi_j=0} \frac{1}{x_j} \right) |D^\chi f(x)|\, dx$$

will appear in the estimates. They will come from the mixture of the leading and remainder terms in separate variables. The term estimated in the proof of Theorem 7.5 is the only one in (7.3) (except the first line) where such a mixture does not appear.

7.2 Equalities

The main part of the preceding one-dimensional activity is almost complete. First, we are now in a position to formulate and prove a precise extension of Theorem 4.15.

Theorem 7.7. *Let* $f : \mathbb{R}^n_+ \to \mathbb{C}$ *be of bounded Hardy variation on* $\mathbb{R}^n_+$ *and let* f *vanish at infinity along with all* $D^\eta f$ *except* $\eta = \mathbf{1}$. *Let also* f *be locally absolutely continuous on* $\mathbb{R}^n_+ \setminus \{0\}$. *In addition, let, for all the derivatives, the inverse formula for the Hilbert transform (cf. (4.16) or (4.21)) hold almost everywhere in each variable. Then*

$$(-1)^{|\eta|}\widehat{f_\eta}(x) = \left(\prod_{j:\eta_j=0} \frac{1}{x_j} \right) B_s^\eta D^\eta f\left(x_\eta, \frac{\pi}{2}x_{\eta-\mathbf{1}}\right)$$

$$+ \sum_{\substack{\chi:\chi_i=0\ if\ \eta_i=1,\\ \chi\neq \mathbf{0},\mathbf{1}}} \left(\prod_{j:\chi_j=1} \frac{1}{x_j} \right)$$

$$\times \prod_{j:(\mathbf{1}-\eta-\chi)_j=1} \left(B_s^j \mathcal{H}_o^j \frac{\partial}{\partial x_j} + B_S^j \frac{\partial}{\partial x_j} \right)$$

$$\times B_s^\eta D^\eta f\left(x_{\mathbf{1}-\chi}, \frac{\pi}{2}x_{-\chi}\right) \qquad (7.8)$$

$$+ \prod_{j:(\mathbf{1}-\eta)_j=1} \left(B_s^j \mathcal{H}_o^j \frac{\partial}{\partial x_j} + B_S^j \frac{\partial}{\partial x_j} \right) B_s^\eta D^\eta f(x).$$

Proof. In fact, the proof is the application of either (4.17) or (4.24) in each variable. It is convenient to apply $|\eta|$ times (4.17) to (6.3) first. By this, we get

$$\widehat{f_\eta}(x) = (-1)^{|\eta|} \int_{\mathbb{R}^{n-|\eta|}_+} B_S^\eta D^\eta f(x_\eta, u_{\mathbf{1}-\eta}) \prod_{i:\eta_i=0} \sin x_i u_i\, du_{\mathbf{1}-\eta}.$$

7.2. Equalities

We then apply (4.24) in each of the variables in the remaining pure sine Fourier transform. Observe that the cases where $\chi = \mathbf{0}, \mathbf{1}$ are written separately in (7.8). The proof is complete. □

One can see that the point is not the proof of Theorem 7.7, it is just the superposition of one-dimensional results. The latter becomes possible due to utilizing the operator B_φ.

The last term in (7.8) can be made the (integrable) remainder term by assuming $D^1 f$ to belong to the product Hardy space. And in general it is the only such a remainder term. The rest of the terms are various types of the leading terms. One cannot get rid of them if one wishes to stay in the most general setting of all functions of bounded Hardy variation. They or some of them disappear (except for the first one, of course), or, more precisely, become of remainder type, if one restricts oneself to certain subspaces of the space of functions of bounded Hardy variation, as in the previous chapter.

The following precise extension of Theorem 4.20, a counterpart of the above theorem, is obtained in the same way.

Theorem 7.9. *Let $f : \mathbb{R}^n_+ \to \mathbb{C}$ be of bounded Hardy variation on $\mathbb{R}^n_+$ and let f vanish at infinity along with all $D^\eta f$ except $\eta = \mathbf{1}$. Let also f and the same $D^\eta f$ be locally absolutely continuous. In addition, let for all the derivatives the inverse formula for the Hilbert transform hold almost everywhere in each variable. Then*

$$\widehat{f}_\eta(x) = B_L^\eta B_c^{1-\eta} D^1 f(x)$$
$$+ \sum_{\substack{\chi : \chi_i = 0 \ if \ \eta_i = 0, \\ \chi \neq \mathbf{0}, \mathbf{1}}} B_L^\chi \prod_{j:(\eta-\chi)_j = 1} \left(B_c^j \mathcal{H}_e^j \frac{\partial}{\partial x_j} + B_C^j \frac{\partial}{\partial x_j} \right) B_c^{1-\eta} D^{1-\eta+\chi} f(x)$$
$$+ \prod_{j:\eta_j = 1} \left(B_c^j \mathcal{H}_e^j \frac{\partial}{\partial x_j} + B_C^j \frac{\partial}{\partial x_j} \right) B_c^{1-\eta} D^{1-\eta} f(x). \tag{7.10}$$

Further, we are also in a position to formulate and prove the result that generalizes Theorem 3.5. All three theorems of this section are proved in a similar way. We will give more details while proving the next one.

Theorem 7.11. *Let $f : \mathbb{R}^n_+ \to \mathbb{C}$ be of bounded Hardy variation on $\mathbb{R}^n_+$ and let f vanish at infinity along with all $D^\eta f$ except $\eta = \mathbf{1}$. Let also f be locally absolutely continuous on $\mathbb{R}^n_+ \setminus \{0\}$. For (6.3), with $\eta = \mathbf{0}$,*

$$\widehat{f_{\mathbf{0}}}(x) = \prod_{j=1}^{n} \frac{1}{x_j} f\left(\frac{\pi}{2x_1}, \ldots, \frac{\pi}{2x_n}\right)$$

$$+ \sum_{\eta \neq \mathbf{0},\mathbf{1}} \left(\prod_{j:\eta_j=1} \frac{1}{x_j}\right) B_G^{1-\eta} D^{1-\eta} f\left(\frac{\pi}{2} x_{-\eta}, x_{\mathbf{1}-\eta}\right)$$

$$+ \sum_{\eta \neq \mathbf{0},\mathbf{1}} (-1)^{n-|\eta|} \left(\prod_{j:\eta_j=1} \frac{1}{x_j}\right) \mathcal{H}_o^{1-\eta} B_s^{1-\eta} D^{1-\eta} f\left(\frac{\pi}{2} x_{-\eta}, x_{\mathbf{1}-\eta}\right)$$

$$+ \sum_{\eta \neq \mathbf{0},\mathbf{1}} (-1)^{n-|\eta|} \mathcal{H}_o^{1-\eta} B_s^{1-\eta} B_G^{\eta} D^1 f(x)$$

$$+ \sum_{\substack{\eta \neq \mathbf{0},\mathbf{1}; \chi \neq \mathbf{0},\mathbf{1}-\eta; \\ \chi_j = 0 \text{ if } j:\eta_j=1}} (-1)^{|\chi|} \left(\prod_{j:\eta_j=1} \frac{1}{x_j}\right) \mathcal{H}_o^{\chi} B_s^{\chi} B_G^{1-\eta-\chi} D^{1-\eta} f\left(\frac{\pi}{2} x_{-\eta}, x_{\mathbf{1}-\eta}\right)$$

$$+ \mathcal{H}_o^{\mathbf{1}} B_s^{\mathbf{1}} D^{\mathbf{1}} f(x) + B_G^{\mathbf{1}} D^{\mathbf{1}} f(x), \tag{7.12}$$

with

$$\int_{\mathbb{R}_+^n} |B_G^{\mathbf{1}} D^{\mathbf{1}} f(x)| \, dx \lesssim \int_{\mathbb{R}_+^n} |D^{\mathbf{1}} f(x)| \, dx. \tag{7.13}$$

In dimension two the fifth line with χ terms does not appear.

Proof. In fact, the proof is the application of (4.24) in each variable, or, in other words, the n times (operator) product of the three terms relation in (4.24). This means that it suffices to verify that all the products do appear in (7.12) in accordance with the chosen notation. Of course, that no extra terms appear should be verified as well. In addition, the way certain terms are grouped in (7.12) should be explained. Let us proceed to this line by line in (7.12). Clearly, the first line presents the "main" leading term to which all (and only) the first terms coming from (4.24) contribute. Similarly, each summand in the last line gives the product of only the second terms in (4.24) and the third ones, respectively.

In the second line we have all possible combinations of the first and third terms in (4.24), in the third line we have all possible combinations of the first and second terms in (4.24), and in the fourth line we have all possible combinations of the second and third terms in (4.24). These lines are relatively simple to write and understand.

More delicate is the fifth line, in which one indicator vector η proves unequal to the task and one more such vector χ is involved. Thus, η indicates those coordinates (j-coordinates with $\eta_j = 1$) for which the first term in (4.24) works. They contribute with the

$$\left(\prod_{j:\eta_j=1} \frac{1}{x_j}\right)$$

7.2. Equalities

factor (as everywhere) and $\frac{\pi}{2}x_{-\eta}$ in the corresponding variables of f. Note that no derivatives of f are taken with respect to these variables. The indicator vector χ characterizes only some of the rest of the variables. We always have $\chi_j = 0$ for the cases where $\eta_j = 1$. Since in the previous lines the cases where only two terms in (4.24) interplay are separated, we exclude the situation $\chi = 1 - \eta$. This means that χ indicates the cases where always both the second and the third terms in (4.24) take part. For such variables, the derivatives of f are always taken; therefore, we have $D^{1-\eta}f$, the derivative of $(n - |\eta|)$-th order. Clearly, the rest of the factors are $\mathcal{H}_o^\chi B_s^\chi$, which always work together and apply here to the χ-th variables with the resulting sign $(-1)^{|\chi|}$, and $B_G^{1-\eta-\chi}$, where B_G applies to each of the $n - |\eta| - |\chi|$ remaining variables. Observe that each summand in the fifth line must have all three different terms. This can never happen in dimension two.

To illustrate this, let us write down the whole formula (7.12) in dimension two, where also it will be much more transparent. With (x, y) in place of (x_1, x_2), it reads as

$$\widehat{f_0}(x,y) = \int_0^\infty \int_0^\infty f(s,t) \sin xs \sin yt \, dsdt = \frac{1}{xy} f\left(\frac{\pi}{2x}, \frac{\pi}{2y}\right)$$
$$+ [\frac{1}{x} B_G^{(0,1)} D^{(0,1)} f(\frac{\pi}{2x}, y) + \frac{1}{y} B_G^{(1,0)} D^{(1,0)} f\left(x, \frac{\pi}{2y}\right)]$$
$$- [\frac{1}{x} \mathcal{H}_o^{(0,1)} B_s^{(0,1)} D^{(0,1)} f(\frac{\pi}{2x}, y) + \frac{1}{y} \mathcal{H}_o^{(1,0)} B_s^{(1,0)} D^{(1,0)} f\left(x, \frac{\pi}{2y}\right)]$$
$$- [\mathcal{H}_o^{(0,1)} B_s^{(0,1)} B_G^{(1,0)} D^{(1,1)} f(x,y) + \mathcal{H}_o^{(1,0)} B_s^{(1,0)} B_G^{(0,1)} D^{(1,1)} f(x,y)]$$
$$+ \mathcal{H}_o^{(1,1)} B_G^{(1,1)} D^{(1,1)} f(x,y) + B_G^{(1,1)} D^{(1,1)} f(x,y). \tag{7.14}$$

This formula almost shows the steps taken for its construction. Indeed, applying (4.24) to the one-dimensional Fourier transform with respect to the second variable, we obtain

$$\int_0^\infty \left[\frac{1}{y} f\left(s, \frac{\pi}{2y}\right) - \mathcal{H}_o B_s D^{(0,1)} f(s,y) + B_G D^{(0,1)} f(s,y)\right] \sin xs \, ds.$$

This gives us three one-dimensional Fourier transforms with respect to the variable s, each for a different function. We again apply (4.24) to each of them. Rewriting them in a proper order gives (7.14). The possibility to proceed to each variable is provided by the assumptions of the theorem, more precisely, by the boundedness of Hardy's variation and absolute continuity. It remains to observe that integrability of the last term in (7.14) (and similarly in the general (7.12)) follows from Lemma 1.126. To check its assumption is routine.

Let us also give the fifth line in (7.12) for dimension three. Here χ can be one of the three vectors $(1,0,0)$, $(0,1,0)$ and $(0,0,1)$, and $(-1)^{|\chi|} = -1$. The same are η-s but always different from the corresponding χ. For example, for $\chi = (1,0,0)$, the indicator vector η can take the values $(0,1,0)$ and $(0,0,1)$ only. Therefore, we have the following six summands, in the variables (x, y, z):

$$-\frac{1}{x}\mathcal{H}_o^{(0,1,0)} B_s^{(0,1,0)} B_G^{(0,0,1)} D^{(0,1,1)} f\left(\frac{\pi}{2x}, y, z\right)$$

$$-\frac{1}{x}\mathcal{H}_o^{(0,0,1)} B_s^{(0,0,1)} B_G^{(0,1,0)} D^{(0,1,1)} f\left(\frac{\pi}{2x}, y, z\right)$$

$$-\frac{1}{y}\mathcal{H}_o^{(1,0,0)} B_s^{(1,0,0)} B_G^{(0,0,1)} D^{(1,0,1)} f\left(x, \frac{\pi}{2y}, z\right)$$

$$-\frac{1}{y}\mathcal{H}_o^{(0,0,1)} B_s^{(0,0,1)} B_G^{(1,0,0)} D^{(1,0,1)} f\left(x, \frac{\pi}{2y}, z\right)$$

$$-\frac{1}{z}\mathcal{H}_o^{(0,1,0)} B_s^{(0,1,0)} B_G^{(1,0,0)} D^{(1,1,0)} f\left(x, y, \frac{\pi}{2z}\right)$$

$$-\frac{1}{z}\mathcal{H}_o^{(1,0,0)} B_s^{(1,0,0)} B_G^{(0,1,0)} D^{(1,1,0)} f\left(x, y, \frac{\pi}{2z}\right).$$

The proof is complete. □

7.2.1 (Even) more general cases

Concerning Theorem 7.11 and its relation with more general cases of the mixed, cosine and sine, transforms, recall that in terms of B_φ operators, the cosine Fourier transform can also be represented to be of the asymptotic form

$$\widehat{f}_c(x) = \int_0^\infty f(t)\cos xt\, dt = -B_s f'(x);$$

see (4.17), which is just integration by parts.

This is meaningful, since the following Hardy type space is meaningful: the one having integrable functions g, for which both $B_s g$ and $\mathcal{H}_o B_s$ are integrable. It is proved above (Subsection 3.3.1 in Chapter 3, see also [121]) that this space is wider than the usual Hardy space for odd functions.

In the previous section we dealt with (6.3), provided $\eta = \mathbf{0}$. Applying (4.17) to (6.3) in the variables for which $\eta_j = 1$ and then using (7.12) for the rest of the variables, we can obtain the following most general result.

Theorem 7.15. *Let $f : \mathbb{R}_+^n \to \mathbb{C}$ be of bounded Hardy variation on $\mathbb{R}_+^n$ and f vanishes at infinity along with all $D^\eta f$ except $\eta = \mathbf{1}$. Let also f be locally absolutely continuous on $\mathbb{R}_+^n \setminus \{0\}$. For (6.3), with $\eta \neq \mathbf{0}, \mathbf{1}$,*

7.2. Equalities

$$\widehat{f_\eta}(x) = (-1)^{|\eta|} \left(\prod_{j:\eta_j=0} \frac{1}{x_j} \right) B_s^\eta D^\eta f \left(x_\eta, \frac{\pi}{2} x_{\eta-1} \right)$$

$$+ (-1)^{|\eta|} \sum_{\substack{\chi \neq \eta, 1-\eta \\ \chi_j = 0 \text{ if } j:\eta_j=1}} \left(\prod_{j:\chi_j=1} \frac{1}{x_j} \right)$$

$$\times B_G^{1-\eta-\chi} D^{1-\chi} B_s^\eta f \left(x_\eta, \frac{\pi}{2} x_{-\chi}, x_{1-\eta-\chi} \right)$$

$$+ \sum_{\substack{\chi \neq \eta, 1-\eta \\ \chi_j = 0 \text{ if } j:\eta_j=1}} (-1)^{n-|\chi|} \left(\prod_{j:\chi_j=1} \frac{1}{x_j} \right)$$

$$\times \mathcal{H}_o^{1-\eta-\chi} B_s^{1-\chi} D^{1-\chi} f \left(x_\eta, \frac{\pi}{2} x_{-\chi}, x_{1-\eta-\chi} \right)$$

$$+ \sum_{\substack{\chi \neq \eta, 1-\eta \\ \chi_j = 0 \text{ if } j:\eta_j=1}} (-1)^{n-|\chi|} \mathcal{H}_o^{1-\eta-\chi} B_s^{1-\chi} B_G^\chi D^1 f(x)$$

$$+ (-1)^{|\eta|} \sum_{\substack{\chi \neq \eta, 1-\eta; \chi_j = 0 \text{ if } j:\eta_j=1 \\ \zeta \neq 1-\eta, 1-\eta-\chi; \zeta_j = 0 \text{ if } j:\chi_j=1}} (-1)^{|\zeta|} \left(\prod_{j:\chi_j=1} \frac{1}{x_j} \right) \mathcal{H}_o^\zeta B_s^{\eta+\zeta} B_G^{1-\eta-\chi-\zeta}$$

$$\times D^{1-\chi} f \left(x_\eta, \frac{\pi}{2} x_{-\chi}, x_{1-\eta-\chi} \right) \qquad (7.16)$$

$$+ (-1)^{|\eta|} \mathcal{H}_o^{1-\eta} B_s^1 D^1 f(x) + (-1)^{|\eta|} B_G^{1-\eta} B_s^\eta D^1 f(x),$$

with

$$\int_{\mathbb{R}_+^n} |B_G^{1-\eta} D^1 B_s^\eta f(x)| \, dx \lesssim \int_{\mathbb{R}_+^n} |D^1 f(x)| \, dx.$$

In dimension three, the terms where ζ is involved do not appear. In dimension two, only the terms in the first and the last lines are present.

For $\eta = \mathbf{1}$, we have

$$\widehat{f_1} = (-1)^n B_s^1 D^1 f(x).$$

For example, in dimension two,

$$\widehat{f_{(1,0)}}(x,y) = \int_0^\infty \int_0^\infty f(s,t) \cos xs \sin yt \, ds \, dt$$

$$= -\frac{1}{y} B_s^{(1,0)} D^{(1,0)} f\left(x, \frac{\pi}{2y}\right) + \mathcal{H}_0^{(0,1)} B_s^{(1,1)} D^{(1,1)} f(x,y)$$

$$- B_G^{(0,1)} B_s^{(1,0)} D^{(1,1)} f(x,y).$$

Of course, if we formally take $\eta = \mathbf{0}$, formula (7.16) reduces to (7.12), while for $\eta = \mathbf{1}$ it reduces to the last relation in Theorem 7.15. In this sense, (7.16) is the most general asymptotic formula for the Fourier transform of an *arbitrary* locally absolutely continuous function with bounded Hardy variation.

7.2.2 The most general situation

If dimension is high enough, $n \geq 4$, one can imagine a combination of all the four opportunities one faces in Theorems 4.15 and 4.20. The general formulation is too superfluous; therefore to get the flavor of such a situation, we proceed to a four-dimensional version, where each of the above phenomena appear as one-dimensional.

Of course, the assumptions are the same, that is, we deal with $f : \mathbb{R}^4_+ \to \mathbb{C}$ of bounded Hardy variation on $\mathbb{R}^4_+$ and vanishing at infinity along with all $D^\eta f$ except $\eta = 1$. Let also f be locally absolutely continuous on $\mathbb{R}^4_+ \setminus \{0\}$. In addition, let for all the derivatives the inverse formula for the Hilbert transform hold almost everywhere in each variable. We present each of the four steps for the asymptotic representation of the corresponding Fourier transform separately as follows:

$$\widehat{f}_{(1,1,0,0)}(x) = \int_{\mathbb{R}^4_+} f(u) \cos x_1 u_1 \cos x_2 u_2 \sin x_3 u_3 \sin x_4 u_4 \, du$$

$$= -\int_{\mathbb{R}^3_+} B^1_s \frac{\partial}{\partial x_1} f(x_1, u_2, u_3, u_4) \cos x_2 u_2 \sin x_3 u_3 \sin x_4 u_4 \, du_2 du_3 du_4$$

$$= -\int_{\mathbb{R}^2_+} B^1_s B^3_c \frac{\partial^2}{\partial x_1 \partial x_3} f(x_1, u_2, x_3, u_4) \cos x_2 u_2 \sin x_4 u_4 \, du_2 du_4$$

$$= -\frac{1}{x_4} \int_{\mathbb{R}_+} B^1_s B^3_c \frac{\partial^2}{\partial x_1 \partial x_3} f\left(x_1, u_2, x_3, \frac{\pi}{2x_4}\right) \cos x_2 u_2 \, du_2$$

$$- \int_{\mathbb{R}_+} B^1_s B^3_c B^4_s \mathcal{H}^4_0 \frac{\partial^3}{\partial x_1 \partial x_3 \partial x_4} f(x_1, u_2, x_3, x_4) \cos x_2 u_2 \, du_2$$

$$- \int_{\mathbb{R}_+} B^1_s B^3_c B^4_S \frac{\partial^3}{\partial x_1 \partial x_3 \partial x_4} f(x_1, u_2, x_3, x_4) \cos x_2 u_2 \, du_2,$$

and finally we get

$$\widehat{f}_{(1,1,0,0)}(x) = \int_{\mathbb{R}^4_+} f(u) \cos x_1 u_1 \cos x_2 u_2 \sin x_3 u_3 \sin x_4 u_4 \, du$$

$$= -\frac{1}{x_4} B^2_L B^1_s B^3_c \frac{\partial^3}{\partial x_1 \partial x_2 \partial x_3} f\left(x_1, x_2, x_3, \frac{\pi}{2x_4}\right)$$

$$- \frac{1}{x_4} B^2_c \mathcal{H}^2_e B^1_s B^3_c \frac{\partial^3}{\partial x_1 \partial x_2 \partial x_3} f\left(x_1, x_2, x_3, \frac{\pi}{2x_4}\right)$$

$$- \frac{1}{x_4} B^2_C B^1_s B^3_c \frac{\partial^3}{\partial x_1 \partial x_2 \partial x_3} f\left(x_1, x_2, x_3, \frac{\pi}{2x_4}\right)$$

$$- B^2_L B^1_s B^3_c B^4_s \mathcal{H}^4_o D^1 f(x) - B^2_c \mathcal{H}^2_e B^1_s B^3_c B^4_s \mathcal{H}^4_o D^1 f(x)$$

$$- B^2_C B^1_s B^3_c B^4_s \mathcal{H}^4_o D^1 f(x) - B^2_L B^1_s B^3_c B^4_S D^1 f(x)$$

$$- B^2_c \mathcal{H}^2_e B^1_s B^3_c B^4_S D^1 f(x) - B^2_C B^1_s B^3_c B^4_S D^1 f(x).$$

7.2. Equalities

The last formula can be made shorter in two ways. First, we know from the one-dimensional theory that the terms where the Hilbert transform appears are integrable by assuming that these Hilbert transforms are integrable, or, in other words, the corresponding derivative belongs to a Hardy space. Under such assumption, these terms may be meant as a function integrable over $\mathbb{R}_+^4$. Second, some other terms may be claimed to be integrable, which is a very special assumption that makes the asymptotic formula rough to a certain extent.

It is now easier to perceive the idea of the existence of a general formula, where each of the four situations will be of arbitrary dimension. For this, all three zero-one vectors η, χ and ζ are needed. Moreover, we will see that an additional vector of this type is needed. Already this is an obvious hint that to try to write down a general formula would be an unrewarding task. Instead, we outline possible steps.

Of course, the initial assumptions are standard: let $f : \mathbb{R}_+^n \to \mathbb{C}$ be of bounded Hardy variation on $\mathbb{R}_+^n$ and f vanishes at infinity along with all $D^\eta f$ except $\eta = \mathbf{1}$. Let also f and the same $D^\eta f$ be locally absolutely continuous. Let us represent the Fourier transform as

$$\widehat{f_\eta}(x) = \widehat{f_\eta}(x_\chi, x_{\eta-\chi}, x_\zeta, x_{\mathbf{1}-\eta-\zeta}),$$

where $\chi \leq \eta$ and $\zeta \leq \mathbf{1} - \eta$, that is, $\chi_j = 0$ if $\eta_j = 0$ and $\zeta_i = 0$ if $\eta_i = 1$. Unfortunately (for notation only, of course) χ and ζ are independent. Here χ indicates the variables with respect to which the cosine Fourier transform is taken and they are subject to conditions related to $\mathcal{H}_o$, while $\eta - \chi$ indicates the variables with respect to which the cosine Fourier transform is taken and they are subject to conditions related to $\mathcal{H}_e$. Similarly, ζ indicates the variables with respect to which the sine Fourier transform is taken and they are subject to conditions related to $\mathcal{H}_o$, while $\mathbf{1} - \eta - \zeta$ indicates the variables with respect to which the sine Fourier transform is taken and they are subject to conditions related to $\mathcal{H}_e$.

In addition, one should assume that for all the derivatives with respect to $x_{\eta-\chi}$ and x_ζ the corresponding inverse formula for the Hilbert transform holds almost everywhere in each of these variables.

The first two steps are relatively transparent. Starting with x_χ and integrating by parts, we obtain

$$\widehat{f_\eta}(x) = (-1)^{|\chi|} B_s^\chi D^\chi (F_{\mathbf{1}-\chi} f)_{\eta-\chi}(x_\chi, x_{\eta-\chi}, x_\zeta, x_{\mathbf{1}-\eta-\zeta}).$$

The next step is very similar and concerns the variables indicated by $\mathbf{1} - \eta - \zeta$. Integrating by parts on the right-hand side of the last relation, we get

$$\widehat{f_\eta}(x) = (-1)^{|\chi|} B_s^\chi B_c^{\mathbf{1}-\eta-\zeta} D^{\chi+\mathbf{1}-\eta-\zeta} (F_{\eta+\zeta-\chi} f)_{\eta-\chi+\zeta}(x_\chi, x_{\eta-\chi}, x_\zeta, x_{\mathbf{1}-\eta-\zeta}).$$

We already know that "Unbearable Lightness Of Being" occurs only in the variables, where no real asymptotics appears. Even denoting, say,

$$\Phi(x_{\eta-\chi}, x_\zeta)$$
$$= (-1)^{|\chi|} B_s^\chi B_c^{\mathbf{1}-\eta-\zeta} D^{\chi+\mathbf{1}-\eta-\zeta} (F_{\eta+\zeta-\chi} f)_{\eta-\chi+\zeta}(x_\chi, x_{\eta-\chi}, x_\zeta, x_{\mathbf{1}-\eta-\zeta}),$$

we see that applying (7.8) to x_ζ and then (7.10) to $x_{\eta-\chi}$ leads to very tedious formulas. For instance, we need at least one additional indicator vector for this. We omit this process. Using the obtained theorems in concrete situations, like above in the four-dimensional case, is a much more clear procedure. Sometimes extra generality only hurts the interests of clarity and applicability.

There is a very good way to get a flavor of such multidimensional results, we have already made use of. It is to consider a function of several variables which is the product of functions of one variable, each satisfying the assumptions of the corresponding one-dimensional theorem. In such a case combining these one-dimensional assumptions is equivalent to the assumptions of a multivariate theorem. Hence, the multidimensional relation will be just the sum of products of functions. But the general case does not differ much in appearance! It is also the sum of of products of operators applied to the corresponding variables rather than the products of functions. This similarity is due to the proper one-dimensional background.

7.3 Szökefalvi-Nagy type theorem

In dimension one, for a function with bounded variation, its derivative exists almost everywhere and is integrable. As we have seen, in several dimensions the situation is very similar. In the above results, to ensure the integrability of the Fourier transform, we assume that the derivative belongs to some space. It was natural to introduce the notation V^X with the meaning $V^X = BV$ if $X = L^1$ and V^X in other cases means that $X \subset L^1$ and $f' \in X$. In particular, V_0^* means that f has bounded variation, vanishes at infinity and $f' \in L^*$.

In several dimensions, for the general Fourier transform over $\mathbb{R}^n$, a similar notation $V^H := V^H(\mathbb{R} \times \cdots \times \mathbb{R})$ is natural, assuming that the right-hand side of (6.8) is finite. Roughly speaking, integrability of the derivatives and their Hilbert transforms are supplemented with integrability in x_j of the function not differentiated in the j-th variable over x_j. This comes from the sine side of the Fourier transform. The finiteness of the corresponding right-hand sides in Corollaries 6.11 and 6.14 will be denoted by $\|f\|_{V^{O_q}} < \infty$ or $\|f\|_{V^{O_\infty}} < \infty$, respectively, since O_q and O_∞ are the notations of the corresponding spaces to which the derivative belongs, like L^* above. For $f : \mathbb{R}^n \to \mathbb{C}$ from any of these classes, we also assume it to be of bounded Hardy variation on $\mathbb{R}^n$ and to vanish at infinity along with all $D^\eta f$ except $\eta = 1$. Also, f and the same $D^\eta f$ are LAC on $\mathbb{R}^n \setminus \{0\}$ in the above sense.

We are now in a position to formulate the main result of this section. It comes from [120], which, in turn, was a multivariate generalization of Trigib's one-dimensional result in [197] (see Section 4.4).

Theorem 7.17. *Let $\{x^k\}_{k=1}^s$ be the points in $\mathbb{R}^n$. Let each of the functions $f(x+x_k)$, $1 \le k \le s$, admit an extension from a neighborhood of the origin and f itself from a neighborhood of infinity ($|x_j| \ge N$, $j = 1, \ldots, n$) to a function from*

7.3. Szőkefalvi-Nagy type theorem

$V^H(\mathbb{R} \times \cdots \times \mathbb{R})$. In order that $f \in W_0(\mathbb{R}^n)$ it is necessary and sufficient, for some $\delta > 0$ and $N > \max_{1 \le k \le s} |x^k|$,

$$\sum_{k=1}^{s} \int_{[0,\delta]^n} \prod_{j=1}^{n} \frac{1}{u_j} \left| \sum_{j=0}^{n} (-1)^j \sum_{|\eta|=j} f\left(x_1^k + (-1)^{\eta_1} u_1, \ldots, x_n^k + (-1)^{\eta_n} u_n\right) \right| du$$

$$+ \int_{[N,\infty)^n} \prod_{j=1}^{n} \frac{1}{u_j} \left| \sum_{j=0}^{n} (-1)^j \sum_{|\eta|=j} f\left((-1)^{\eta_1} u_1, \ldots, (-1)^{\eta_n} u_n\right) \right| du < \infty \quad (7.18)$$

to hold.

To be precise, this is a kind of a simplest generalization of Theorem 4.35, with the singularities just points, as in dimension one. However, one may imagine a more complicated type of singularities, say curves or surfaces. Nevertheless, it is not clear how to even formulate such extensions.

7.3.1 Auxiliary results

We need some auxiliary results. As the reader may remember, one of them in its one-dimensional form is already used in the proof of Theorem 4.35.

Lemma 7.19. *Let each function h_j, $j = 1, \ldots, n$, be with compact support and belong to* Lip1. *If $g \in V^H(\mathbb{R} \times \cdots \times \mathbb{R})$, then also*

$$H(u)g(u) = \prod_{j=1}^{n} h(u_j)g(u) \in V^H(\mathbb{R} \times \cdots \times \mathbb{R}).$$

Proof. Again, the proof in dimension two will give a clear view of the general situation. We have

$$\int_{\mathbb{R}^2} \frac{h_1(u)h_2(v)g(u,v)}{(x-u)(y-v)} \, du \, dv$$

$$= \int_{\mathbb{R}^2} \frac{h_1(u)[h_2(v) - h_2(y)]g(u,v)}{(x-u)(y-v)} \, du \, dv$$

$$+ h_2(y) \int_{\mathbb{R}^2} \frac{h_1(u)g(u,v)}{(x-u)(y-v)} \, du \, dv$$

$$= \int_{\mathbb{R}^2} \frac{[h_1(u) - h_1(x)][h_2(v) - h_2(y)]g(u,v)}{(x-u)(y-v)} \, du \, dv$$

$$+ h_1(x) \int_{\mathbb{R}^2} \frac{[h_2(v) - h_2(y)]g(u,v)}{(x-u)(y-v)} \, du \, dv$$

$$+ h_2(y) \int_{\mathbb{R}^2} \frac{[h_1(u) - h_1(x)]g(u,v)}{(x-u)(y-v)} \, du \, dv$$

$$+ h_1(x)h_2(y) \int_{\mathbb{R}^2} \frac{g(u,v)}{(x-u)(y-v)} \, du \, dv.$$

The L^1 norm of the first integral on the right is estimated as

$$\int_{\mathbb{R}^2} |g(u,v)| \int_{\mathbb{R}} \frac{|h_1(u) - h_1(x)|}{|x-u|} dx \int_{\mathbb{R}} \frac{|h_2(v) - h_2(y)|}{|y-v|} dy \, du \, dv$$
$$\leq C \int_{\mathbb{R}^2} |g(u,v)| \, du \, dv.$$

The last one is simply bounded by the L^1 norm of $\mathcal{H}_{(1,1)}$. The intermediate two are estimated in a similar way, let us show it for the first one. We get

$$\int_{\mathbb{R}^2} \left| \int_{\mathbb{R}^2} \frac{[h_2(v) - h_2(y)]g(u,v)}{(x-u)(y-v)} du \, dv \right| dx \, dy$$
$$\leq \int_{\mathbb{R}} dx \int_{\mathbb{R}} dy \int_{\mathbb{R}} \left| \int_{\mathbb{R}} \frac{g(u,v)}{(x-u)} du \right| \frac{|h_2(v) - h_2(y)|}{|y-v|} dv$$
$$= \int_{\mathbb{R}} dx \int_{\mathbb{R}} dv \left| \int_{\mathbb{R}} \frac{g(u,v)}{(x-u)} du \right| \int_{\mathbb{R}} \frac{|h_2(v) - h_2(y)|}{|y-v|} dy$$
$$\leq C \int_{\mathbb{R}^2} \left| \int_{\mathbb{R}} \frac{g(u,v)}{(x-u)} du \right| dx \, dv.$$

The proof is complete. $\square$

Remark 7.20. Of course, one can take a function $H(u)$ with, say, all the derivatives $D^\eta H$ bounded instead of $\prod_{j=1}^n h(u_j)$ with Lip1 functions h_j.

Lemma 7.21. *Let $f : \mathbb{R}^n \to \mathbb{C}$ be of bounded Hardy variation on $\mathbb{R}^n$ and f vanishes at infinity along with all $D^\eta f$ except $\eta = \mathbf{1}$. Let also f and the same $D^\eta f$ be locally absolutely continuous on $\mathbb{R}^n \setminus \{0\}$ in the above sense. If $f(\cdot + x^0) \in V^H(\mathbb{R} \times \cdots \times \mathbb{R})$, then*

$$\left| \int_{\substack{a_j \leq |x_j| \leq b_j \\ j=1,\ldots,n}} |\widehat{f}(x)| \, dx - 2^n \int_{[\frac{\pi}{2b_1}, \frac{\pi}{2a_1}] \times \cdots \times [\frac{\pi}{2b_n}, \frac{\pi}{2a_n}]} \left(\prod_{j=1}^n \frac{1}{u_j} \right) \right.$$
$$\left. \times \left| \sum_{j=0}^n (-1)^j \sum_{|\eta|=j} f\left(x_1^0 + (-1)^{\eta_1} u_1, \ldots, x_n^0 + (-1)^{\eta_n} u_n \right) \right| du \right|$$
$$\lesssim \|f(\cdot + x^0)\|_{V^H}. \tag{7.22}$$

Proof. We have

$$\widehat{f}(x) = e^{-i(x,x^0)} \int_{\mathbb{R}^n} f(u + x^0) e^{-i(x,u)} du.$$

7.3. Székefalvi-Nagy type theorem

Applying Theorem 6.4, we obtain

$$\left| \int_{\substack{a_j \le |x_j| \le b_j \\ j=1,\ldots,n}} |\widehat{f}(x)|\, dx - \int_{\substack{a_j \le |x_j| \le b_j \\ j=1,\ldots,n}} \left(\prod_{j=1}^{n} \frac{1}{|u_j|} \right) \right.$$
$$\left. \times \left| \sum_{j=0}^{n} (-1)^j \sum_{|\eta|=j} f\left(x_1^0 + (-1)^{\eta_1} \frac{\pi}{2|u_1|}, \ldots, x_n^0 + (-1)^{\eta_n} \frac{\pi}{2|u_n|} \right) \right| du \right|$$
$$\lesssim \|f(\cdot + x^0)\|_{V^H}.$$

The change of variables in the second integral $\frac{\pi}{2|u_j|} \to u_j$, $j=1,\ldots,n$, completes the proof. $\square$

7.3.2 Proof of Theorem 7.17

It is well known that for a function with bounded Hardy variation, the Fourier transform exists as an improper integral, see, e.g., [198, 3.3.10].

Setting

$$\delta = \frac{1}{3} \min_{k \ne m, j=1,\ldots,n} |x_j^k - x_j^m|,$$

we assume that each of the functions $f(x^k + \cdot)$ can be extended from the cubic 2δ-neighborhood of the origin to a function from V^H. Let H be such as defined in Lemma 7.19, with $H(x) = 1$ in the cubic δ-neighborhood of the origin. Then, by Lemma 7.19, the function

$$f_k(x) = H(x - x^k) f(x) \in V^H.$$

Using Lemma 7.21 and the fact that $H(x) = 1$ in the cubic δ-neighborhood of the origin, we have $f_k \in W_0$ if and only if the k-th summand in the sum in (7.18) is finite. This is true for every k, $1 \le k \le s$.

Let now

$$f(x) = \sum_{k=1}^{s} f_k(x) + f_{s+1}(x).$$

Here

$$f_{s+1}(x) = f(x)$$

if

$$|x_j| \ge N = 2\delta + \max_{1 \le k \le s} |x^k|, j=1,\ldots,n,$$

while for

$$|x_j| \le N - \delta, \quad j=1,\ldots,n,$$

we have $f_{s+1} \in \text{Lip1}$ in each variable uniformly with respect to the other ones. Similarly to what has been done above, we set

$$H_{s+1}(x) = \begin{cases} 1, & |x_j| \geq N, \quad j = 1, \ldots, n, \\ 0, & |x_j| \leq N - \delta, \quad j = 1, \ldots, n, \\ \text{linear}, & \text{otherwise.} \end{cases}$$

Therefore,

$$\begin{aligned} f_{s+1}(x) &= f_{s+1}(x) H_{s+1}(x) + f_{s+1}(x)(1 - H_{s+1}(x)) \\ &= J_1(x) + J_2(x). \end{aligned} \tag{7.23}$$

The second term J_2 belongs to W_0 because of Lemma 7.19 and Remark 7.20. Since all the f_k and f belong to V^H, we get $f_{s+1} \in V^H$. Hence, by Lemma 7.21, the first term J_1 on the right-hand side of (7.23) belongs to W_0 if and only if

$$\int_{[N-\delta,\infty)^n} \prod_{j=1}^n \frac{1}{u_j} \left| \sum_{j=0}^n (-1)^j \sum_{|\eta|=j} J_1\left((-1)^{\eta_1} u_1, \ldots, (-1)^{\eta_n} u_n\right) \right| du < \infty,$$

which is equivalent to the finiteness of the last term in (7.18). The proof is complete. $\square$

Chapter 8
Bounded variation and sampling

In this chapter, we shall consider certain problems where a function of bounded variation generates trigonometric series which are then compared with its Fourier integral. In fact, Chapter 4 in [198] is devoted to these problems. Here, the more modern term "sampling" is equivalent to the older "discretization". Those who expected a sort of Whittaker–Kotel'nikov–Shannon type matter might be disappointed. The present chapter can be considered as a development of certain of the results in [198]. One type of these results, the Poisson summation formula, is old and classical. However, results related to bounded variation are specific and a bit off a field. The other one is more recent and aims to the comparison of Fourier integrals and trigonometric series related to functions and sequences with bounded variation. We start from this type of problems.

It will differ in certain respects from the previous matter. First of all, we shall back away from the intention to separate one-dimensional and multidimensional results. Some one-dimensional assertions and proofs will appear here instead of being presented in the first part.

8.1 Bridge

The main problem in the integrability of trigonometric series reads as follows. We begin with the one-dimensional setting. Given a trigonometric series

$$\frac{a_0}{2} + \sum_{k=1}^{\infty} (a_k \cos kx + b_k \sin kx),$$

find assumptions on the sequences of coefficients $\{a_k\}, \{b_k\}$ under which the series is the Fourier series of an integrable function. Frequently, the series

$$\frac{a_0}{2} + \sum_{k=1}^{\infty} a_k \cos kx \qquad (8.1)$$

and
$$\sum_{k=1}^{\infty} b_k \sin kx \qquad (8.2)$$
are investigated separately, since there is a difference in their behavior. Usually, integrability of (8.2) requires additional assumptions. However, one of the basic assumptions for both (8.1) and (8.2) is that the sequence $\{a_k\}$ or $\{b_k\}$ is of bounded variation, written $\{a_k\} \in bv$ or $\{b_k\} \in bv$, that is, satisfies the condition
$$\sum_{k=1}^{\infty} |\Delta a_k| < \infty,$$
where $\Delta a_k = a_k - a_{k+1}$, and similarly for Δb_k.

The history of the topic goes back to more than a century. After decades of development of the topic, two peaks were achieved. One is the so-called Boas–Telyakovskii type condition, see, e.g., [24], [16], [181, 182, 183]. Theorem 2.8 is a direct extension of these results to the non-periodic case. The other peak were the results in [10] for amalgam type spaces. They were generalized for Fourier transforms in [115]. Of course, there were many other results, some more and some less related; a sort of a survey can be found in [112].

However, one of the main features of the generalizations to the Fourier transforms was that they allowed one to derive the known and new results for the integrability of trigonometric series "for free". Moreover, as an application of integrability results for the Fourier transform, one could obtain integrability results for (8.2) even in a stronger, asymptotic form, see [107].

We can relate this problem on the integrability of trigonometric series to a similar one for Fourier transforms as follows. First, given series (8.1) or (8.2) with the null sequence of coefficients being in an appropriate sequence space, set for $x \in [k, k+1)$,
$$a(x) = a_k + (k-x)\Delta a_k, \quad a_0 = 0,$$
$$b(x) = b_k + (k-x)\Delta b_k.$$

So, we construct a corresponding function by means of linear interpolation of the sequence of coefficients. Of course, one may interpolate not only linearly, but there are no problems where this might be of importance so far.

Further, for functions of bounded variation f, to pass from series to integrals and vice versa, we will make use of a certain type of results that, with certain abuse of terminology, will be called a "bridge". In fact, this is one of the possible ways of discretization, well adjusted to bounded variation. Of course, such a bridge is usable in both directions but we emphasize the passage from trigonometric series to the Fourier transform, for which the obtained results are applicable.

We study the trigonometric series
$$\sum_{k \in \mathbb{Z}_+^n} a_k \prod_{i:\eta_i=1} \cos x_i k_i \prod_{i:\eta_i=0} \sin x_i k_i, \qquad (8.3)$$

8.1. Bridge

with the null sequence of coefficients a_k being in an appropriate sequence space, set for $k \in [k_1, k_1 + 1) \times \cdots \times [k_n, k_n + 1)$,

$$A(x) = \sum_{0 \leq \eta \leq 1} \prod_{j:\eta_j=1} (k_j - x_j) \Delta_\eta a_k, \tag{8.4}$$

where

$$\Delta_\eta a_k = \prod_{j:\eta_j=1} (a_{k_1,\ldots,k_{j-1},k_j,k_{j+1},\ldots,k_n} - a_{k_1,\ldots,k_{j-1},k_j+1,k_{j+1},\ldots,k_n}).$$

In the multivariate case, it is natural to consider trigonometric series with the sequence of coefficients of bounded Hardy variation, that is, such that

$$\sum_{k \in \mathbb{Z}_+^n} |\Delta_\eta a_k| < \infty$$

for each $\eta \neq \mathbf{0}$. Indeed, such a trigonometric series converges almost everywhere in the Pringsheim sense (see, e.g., [149]).

8.1.1 One-dimensional bridge

In [18], the function f is assumed to be of bounded variation (and of compact support which is an unnecessary restriction). Further results are due to Trigub. In dimension one, his estimates are of great generality and sharpness, first of all, see [196] and [198, Th.4.1.2]. For the function f of bounded usual (one-dimensional) total variation $Vf := V_\mathbb{R} f$ and vanishing at infinity, it reads as follows:

$$\sup_{|x| \leq \pi} \left| \int_\mathbb{R} f(t) e^{ixt} dt - \sum_{k \in \mathbb{Z}} f(k) e^{ixk} \right| \lesssim \|f\|_{BV}. \tag{8.5}$$

Let us present one of the possible proofs. It will give a flavor of how the problem can be solved and extended to higher dimensions and of possible difficulties using this approach.

Proof. Denoting

$$s(t) = \frac{t}{2 \sin \frac{t}{2}}$$

and using a standard trick (see, e.g., [211, Ch.5, Th.2.29])

$$e^{ikx} = s(x) \int_{k-\frac{1}{2}}^{k+\frac{1}{2}} e^{ixt} dt, \tag{8.6}$$

we get

$$\sum_{k \in \mathbb{Z}} f(k) e^{ixk} = s(x) \sum_{k \in \mathbb{Z}} f(k) \int_{k-\frac{1}{2}}^{k+\frac{1}{2}} e^{ixt} dt.$$

Since
$$\int_{\mathbb{R}} f(t)e^{ixt}dt = \sum_{k \in \mathbb{Z}} \int_{k-\frac{1}{2}}^{k+\frac{1}{2}} f(t)e^{ixt}dt,$$
we have
$$\int_{\mathbb{R}} f(t)e^{ixt}dt - \sum_{k \in \mathbb{Z}} f(k)e^{ixk}$$
$$= s(x)\sum_{k \in \mathbb{Z}} \int_{k-\frac{1}{2}}^{k+\frac{1}{2}} [f(t)-f(k)]e^{ixt}dt + (1-s(x))\int_{\mathbb{R}} f(t)e^{ixt}dt.$$

Observe that
$$1 \leq |s(t)| \leq \frac{\pi}{2} \tag{8.7}$$
for $0 < |t| < \pi$. By this, the sum on the right-hand side is immediately bounded by $\|f\|_{BV}$. Taking into account that
$$\frac{s(t)-1}{t}$$
is also bounded, we arrive at
$$x \int_{\mathbb{R}} f(t)e^{ixt}dt,$$
which needs to be checked. Integrating by parts in the Stieltjes sense, we see that
$$-i\int_{\mathbb{R}} e^{ixt}df(t)$$
remains. But it is dominated by $\|f\|_{BV}$, as required. $\square$

To show how the erected bridge is crossed, we give only one very recent result. Its main feature is that it had never been proved directly for the sequence of coefficients but was derived by means of the same new result for the Fourier transform, Theorem 2.20. The relation (8.5) allows us to pass from estimating trigonometric series (8.1) and (8.2) to estimating the Fourier transform of $a(t)$ and $b(t)$, respectively. Earlier results derived from Theorem 2.8 can be found already in [107].

Theorem 8.8. *If the coefficients* $\{a_k\}$ *in* (8.1) *and* $\{b_k\}$ *in* (8.2) *tend to 0 as* $k \to \infty$, *and the sequences* $\{\Delta a_k\}$ *and* $\{\Delta b_k\}$ *are in* h_e, *then* (8.2) *represents an integrable function on* $[0, \pi]$, *and, for* $N \leq \frac{\pi}{2x} < N+1$,

$$\sum_{k=1}^{\infty} a_k \cos kx = \frac{A}{x}a\left(\frac{\pi}{2x}\right) + \frac{\pi}{2x}\sum_{k=1}^{N-1} \Delta a_k \ln \frac{e\pi}{2x(k+1)(1+\frac{1}{k})^k} + G(x), \tag{8.9}$$

8.1. Bridge

where

$$A = \frac{2}{\pi}\left(\int_0^{\frac{\pi}{2}} \frac{1-\cos t}{t}\,dt - \int_{\frac{\pi}{2}}^\infty \frac{\cos t}{t}\,dt\right)$$

and

$$\int_0^\pi |G(x)|\,dx \lesssim \|\{\Delta b_k\}\|_{h_e}.$$

Proof. By (8.5) we can deal with the functions $a(x)$ and $b(x)$ instead of the given sequences. Placing $g(t) = a'(t)$ in $\mathcal{H}_e g$ and taking into account that $a'(t)$ equals $-\Delta a_k$ if $k \leq t < k+1$, we prove the first part of the theorem by fulfilling routine calculations.

We treat the remainder term in (8.9) in the same manner. What remains to prove is that

$$\frac{2}{\pi x}\int_0^{\frac{\pi}{2x}} a'(u)\ln\frac{2ux}{\pi}\,du$$

can be reduced to the second leading term on the right-hand side of (8.9). Indeed, it can be represented as

$$-\frac{2}{\pi x}\sum_{k=0}^{N-1}\Delta a_k \int_k^{k+1} \ln\frac{2ux}{\pi}\,du - \frac{2}{\pi x}\Delta a_N \int_N^{\frac{\pi}{2x}} \ln\frac{2ux}{\pi}\,du. \quad (8.10)$$

The first integral is elementary and is exactly the logarithm in the second leading term on the right-hand side of (8.9). It is worth recalling the classical limit

$$\lim_{k\to\infty}\left(1+\frac{1}{k}\right)^k = e$$

and that the sequence in this limit is increasing. Routine calculations with the first term of the sum and with the second integral in (8.10) complete the proof of the theorem. □

It is worth discussing the nature of the term

$$\frac{1}{x}a\left(\frac{\pi}{2x}\right).$$

We have

$$\int_{\frac{\pi}{2(N+1)}}^{\frac{\pi}{2}} \frac{1}{x}\left|a\left(\frac{\pi}{2x}\right)\right|dx = \sum_{k=1}^N \int_{\frac{\pi}{2(k+1)}}^{\frac{\pi}{2k}} \frac{1}{x}\left|a\left(\frac{\pi}{2x}\right)\right|dx$$

$$= \sum_{k=1}^N \int_k^{k+1} \frac{1}{x}|a(x)|\,dx.$$

The right-hand side is obviously equivalent to

$$\sum_{k=1}^{N} \frac{|a_k|}{k}.$$

We are going to discuss similar operations in several dimensions, that is, to generalize (8.5), dealing with functions and sequences with bounded Hardy variation only.

8.1.2 Temporary bridge

There are certain difficulties in extending (8.5) directly. The recent paper [130], where such a construction was deeply studied, gives both an overview of the preceding efforts and difficulties such efforts face. It has been known for quite a long time that to prove a direct multidimensional version of (8.5) for functions with bounded Hardy variation is impossible. All the attempts in minimizing additional assumptions were reduced to finding an appropriate different variation that the function satisfies. Here we present a version of a bridge constructed only by means of one variation, the Hardy one. Additional conditions do appear of course, but they are more natural within the scope of the integrability of the Fourier transforms.

We decided to divide the process into two steps. At first, we erect a "temporary" construction. It assumes only the boundedness of Hardy's variation but does not allow one to get rid of all the series, some continue to support the construction. This result which we call a "temporary bridge" is

Theorem 8.11. *Let $f(x)$ be a function of bounded Hardy variation on $\mathbb{R}^n$. Let, in addition, VV_η vanish at infinity for all η except $\eta = \mathbf{1}$. Then*

$$\sup_{|x_j|\leq\pi, j=1,2,\ldots,n} \left| \sum_{k\in\mathbb{Z}^n} f(k) e^{i\langle x,k\rangle} \right. \tag{8.12}$$

$$+ \sum_{\eta\neq 0,\mathbf{1}} (-1)^{|\eta|} \sum_{k_\eta} \int_{\mathbb{R}^{n-|\eta|}} f(k_\eta, u_{\mathbf{1}-\eta}) e^{i\langle x_{\mathbf{1}-\eta}, u_{\mathbf{1}-\eta}\rangle} du_{\mathbf{1}-\eta}\, e^{i\langle x_\eta, k_\eta\rangle} \tag{8.13}$$

$$\left. + (-1)^n \int_{\mathbb{R}^n} f(u) e^{i\langle x,u\rangle} du \right| \lesssim \|f\|_{VH(f)}. \tag{8.14}$$

For the sake of simplicity, we formulate and prove the two-dimensional version. The general case is treated in completely the same way.

8.1. Bridge

Theorem 8.15. *Let $f(x,y)$ be a function of bounded Hardy variation on $\mathbb{R}^2$. Let, in addition, $\lim\limits_{|x|+|y|\to\infty} f(x,y), VV_{(1,0)}(y), VV_{(0,1)}(x) = 0$. Then*

$$\sup_{|x|,|y|\leq\pi}\left|\sum_{k,l}f(k,l)e^{i(kx+ly)} - \sum_k\int_{\mathbb{R}}f(k,v)e^{iyv}dv\,e^{ikx}\right.$$
$$\left.-\sum_l\int_{\mathbb{R}}f(u,l)e^{ixu}du\,e^{ily} + \int_{\mathbb{R}^2}f(u,v)e^{i(xu+yv)}du\,dv\right| \lesssim \|f\|_{VH(f)}. \tag{8.16}$$

Proof. With the same

$$s(t) = \frac{t}{2\sin\frac{t}{2}}$$

in hand and using again (8.6), we have

$$\sum_{k,l}f(k,l)e^{i(kx+ly)} - \sum_k\int_{\mathbb{R}}f(k,v)e^{iyv}dv\,e^{ikx}$$
$$= \sum_k s(y)\sum_l\int_{l-\frac{1}{2}}^{l+\frac{1}{2}}[f(k,l)-f(k,v)]e^{iyv}dv\,e^{ikx}$$
$$+ \sum_k[s(y)-1]\int_{\mathbb{R}}f(k,v)e^{iyv}dv\,e^{ikx}. \tag{8.17}$$

We now calculate

$$\sum_{k,l}f(k,l)e^{i(kx+ly)} - \sum_k\int_{\mathbb{R}}f(k,v)e^{iyv}dv\,e^{ikx}$$
$$-\sum_l\int_{\mathbb{R}}f(u,l)e^{ixu}du\,e^{ily} + \int_{\mathbb{R}^2}f(u,v)e^{i(xu+yv)}du\,dv. \tag{8.18}$$

We can represent the two last summands in (8.18) as

$$-s(y)\sum_l\int_{l-\frac{1}{2}}^{l+\frac{1}{2}}[f(u,l)-f(u,v)]e^{i(xu+yv)}du\,dv$$
$$-[s(y)-1]\int_{\mathbb{R}^2}f(u,v)e^{i(xu+yv)}du\,dv.$$

This means that (8.18) is just equivalent to the application of the procedure used in (8.17) in l to the left-hand side of (8.17) but in k. In other words, we can deal with

$$\sum_{k,l} f(k,l)e^{i(kx+ly)} - \sum_k \int_{\mathbb{R}} f(k,v)e^{iyv}dv\, e^{ikx}$$

$$- s(y)\sum_l \int_{l-\frac{1}{2}}^{l+\frac{1}{2}} [f(u,l)-f(u,v)]e^{i(xu+yv)}du\,dv$$

$$- [s(y)-1]\int_{\mathbb{R}^2} f(u,v)e^{i(xu+yv)}du\,dv$$

rather than with (8.18). This value is equal to

$$s(x)s(y)\sum_{k,l}\int_{k-\frac{1}{2}}^{k+\frac{1}{2}}\int_{l-\frac{1}{2}}^{l+\frac{1}{2}} [f(k,l)-f(k,v)$$

$$- f(u,l)+f(u,v)]e^{i(xu+yv)}du\,dv$$

$$+ s(y)[s(x)-1]\sum_l \int_{l-\frac{1}{2}}^{l+\frac{1}{2}}\int_{\mathbb{R}} [f(u,l)-f(u,v)]e^{i(xu+yv)}du\,dv$$

$$+ s(x)[s(y)-1]\sum_k \int_{k-\frac{1}{2}}^{k+\frac{1}{2}}\int_{\mathbb{R}} [f(k,v)-f(u,v)]e^{i(xu+yv)}du\,dv$$

$$+ [s(x)-1][s(y)-1]\int_{\mathbb{R}^2} f(u,v)e^{i(xu+yv)}du\,dv. \qquad (8.19)$$

Inequalities (8.7) immediately imply that the first summand in (8.19) is dominated by $\|f\|_{VH(f)}$. Taking again into account that

$$\frac{s(t)-1}{t}$$

is bounded, we get the same estimate for the last summand in (8.19) by integrating by parts repeatedly in each variable as a Stieltjes integral. The intermediate two summands are treated in the same way, by integrating by parts in x where we have the factor $s(x)-1$ and in y where we have the factor $s(y)-1$. The proof is complete. □

8.1.3 Stable bridge

To be able to apply the results for the Fourier transform to trigonometric series, it is desirable to make use of Theorem 8.11. Its assertions (8.12) and (8.14) are appropriate for this. However, (8.13) is a mixture of sums and integrals, and further calculations are needed in order to get rid of the sums in (8.13). The price for it is twofold. First, we let the function be from a specific function space in addition to the boundedness of Hardy's variation. Such a space is natural in the problems

8.1. Bridge

of integrability of the Fourier transform of a function of bounded Hardy variation. Secondly, we estimate the deviation between the series and the integral in the L^1 norm over $\mathbb{T}^n$, with $\mathbb{T} = [-\pi, \pi)$, rather than in the sup-norm. This is, in a sense, resetting, since even in the first result of that sort by Belinsky, [18], L^p norms were estimated.

Theorem 8.20. *Let f be of bounded Hardy variation and let f vanish at infinity along with all $D^\eta f$ except $\eta = \mathbf{1}$. Let also f and the same $D^\eta f$ be absolutely continuous. Then*

$$\sum_{k \in \mathbb{Z}^n} f(k) e^{i\langle x, k \rangle} = \int_{\mathbb{R}^n} f(u) e^{i\langle x, u \rangle} du + \Phi(x), \qquad (8.21)$$

where

$$\int_{\mathbb{T}^n} |\Phi(x)| \, dx \lesssim \sum_{\eta \neq \mathbf{0}} \int_{\mathbb{R}^{|\eta|}} \left(\prod_{j:\eta_j = 0} \int_{\mathbb{T}} \frac{1}{|x_j|} \right) |F_{\mathbf{1}-\eta}(D^{\mathbf{1}} f)(x_{\mathbf{1}-\eta}, u_\eta)| \, dx_{\mathbf{1}-\eta} \, du_\eta. \qquad (8.22)$$

Proof. Thus, we deal with each of the summands, corresponding to $\eta \neq \mathbf{0}, \mathbf{1}$, in (8.13):

$$I_\eta = \sum_{k_\eta} \int_{\mathbb{R}^{n-|\eta|}} f(k_\eta, u_{\mathbf{1}-\eta}) e^{i\langle x_{\mathbf{1}-\eta}, u_{\mathbf{1}-\eta}\rangle} du_{\mathbf{1}-\eta} \, e^{i\langle x_\eta, k_\eta \rangle}.$$

Observe that the case $\eta = \mathbf{1}$ on the right-hand side of (8.22) comes from the right-hand side of (8.14). We then apply to the $|\eta|$-dimensional sum in I_η the corresponding assertion (8.12)-(8.14) in Lemma 8.11. It will be controlled by Hardy's variation of $F_{\mathbf{1}-\eta} f$ on $\mathbb{R}^{|\eta|}$.

In the integral, let us integrate by parts $(n - |\eta|)$ times, repeatedly in each of the variables u_j for which $\eta_j = 0$. Since integrated terms vanish, we obtain

$$I_\eta = \sum_{k_\eta} \prod_{j:\eta=0} \frac{1}{ix_j} F_{\mathbf{1}-\eta} f(k_\eta, -x_{\mathbf{1}-\eta}) \, e^{i\langle x_\eta, k_\eta \rangle}.$$

The L^1 norm of this value gives a part of the right-hand side in (8.22). It remains to observe that due to the signs

$$\sum_{\eta \neq \mathbf{0}} (-1)^{|\eta|} = -1.$$

Mixed relations of type I_η but of lower dimension are treated in the same manner, which completes the proof. □

Remark 8.23. In fact, this bridge can be even "firmer" if one gets rid of the absolute continuity assumption. The result will be just the same but with $d^{\mathbf{1}}$ in place of $D^{\mathbf{1}}$ on the right-hand side of (8.22).

170   Chapter 8. Bounded variation and sampling

Now, when we have a stable bridge, let us cross it towards results on trigonometric series. For this, we need the discrete versions of the Hilbert transforms (see (1.66)-(1.68) in Chapter 1. Their multivariate extensions go along the same lines as for functions. Indeed, in this case, similarly to that for functions, we can use the notation
$$h^1(\mathbb{Z} \times \cdots \times \mathbb{Z}) := h_n^1(\mathbb{Z} \times \cdots \times \mathbb{Z})$$
and a modified definition on the basis of the discrete Hilbert transforms applied to each variable. Such a transform applied to the j-th variable will be defined $\hbar_j$ and, consequently, $\hbar_j \hbar_k \cdots \hbar_l := \hbar_{jk\ldots l}$. However, there is a more convenient notation for this. We shall define $h_\eta^1(\mathbb{Z} \times \cdots \times \mathbb{Z})$ if the discrete Hilbert transform with respect to each of the j-th variables, for which $\eta_j = 1$, is the even discrete Hilbert transform, while for the rest it is odd. We will be especially interested in the case where $\eta = \mathbf{0}$, that is, all the discrete Hilbert transforms are odd. The corresponding notation is similar to that in dimension one: $h_\mathbf{0}^1(\mathbb{Z} \times \cdots \times \mathbb{Z})$.

By this, the norm of a sequence a in $h^1(\mathbb{Z} \times \cdots \times \mathbb{Z})$ will be

$$\|a\|_{h^1(\mathbb{Z}\times\cdots\times\mathbb{Z})} = \sum_{0 \leq \eta \leq 1} \|\hbar_\eta a\|_{\ell^1(\mathbb{Z}^n)}. \tag{8.24}$$

The result for the trigonometric series (see (8.3))

$$\sum_{k \in \mathbb{Z}_+^n} a_k \prod_{i:\eta_i=1} \cos x_i k_i \prod_{i:\eta_i=0} \sin x_i k_i,$$

with the null sequence of coefficients a_k and with, for $k \in [k_1, k_1+1) \times \cdots \times [k_n, k_n+1)$,

$$A(x) = \sum_{\mathbf{0} \leq \eta \leq 1} \prod_{j:\eta_j=1} (k_j - x_j) \Delta_\eta a_k,$$

see (8.4), reads as follows.

Theorem 8.25. *Let a be a sequence with bounded Hardy's variation on $\mathbb{Z}_+^n$ and let the sequence a vanish at infinity along with all $\Delta_\eta a$ except $\eta = \mathbf{1}$. If $\Delta_\mathbf{1} a \in h_\mathbf{0}^1(\mathbb{Z} \times \cdots \times \mathbb{Z})$, then for $\eta = \mathbf{1}$ we have*

$$\int_{\mathbb{T}_+^n} \left| \sum_{\mathbb{Z}_+^n} a_k \prod_{i=1}^n \cos x_i k_i \right| dx \lesssim \|\Delta_\mathbf{1} a\|_{h_\mathbf{0}^1(\mathbb{Z}\times\cdots\times\mathbb{Z})}; \tag{8.26}$$

for $\eta \neq \mathbf{1}, \mathbf{0}$,

$$\int_{\mathbb{T}_+^n} \left| \sum_{k \in \mathbb{Z}_+^n} a_k \prod_{i:\eta_i=1} \cos x_i k_i \prod_{i:\eta_i=0} \sin x_i k_i \right| dx$$

$$\lesssim \sum_{\substack{\chi \neq \mathbf{1},\mathbf{0} \\ \mathbf{0} \leq \chi \leq \eta}} \sum_{\substack{\mathbf{0} \leq \chi \leq \mathbf{1}-\eta \\ \mathbf{0} \leq \zeta \leq \chi}} \sum_{k \in \mathbb{Z}_+^n} \frac{|\hbar_o^{\chi+\zeta} \Delta_{\eta+\chi} a_k|}{\prod_{j:\chi_j=\eta_j=0}(k_j+1)}; \tag{8.27}$$

8.1. Bridge   171

and for $\eta = \mathbf{0}$,

$$\sum_{k \in \mathbb{Z}_+^n} a_k \prod_{i=1}^n \sin x_i k_i = A\left(\frac{\pi}{2x_1}, \ldots, \frac{\pi}{2x_n}\right) \prod_{j=1}^n \frac{1}{x_j} + F(x), \qquad (8.28)$$

with

$$\int_{\mathbb{T}_+^n} |F(x)|\, dx \lesssim \sum_{\substack{0 \le \chi \le 1, \chi \ne 0 \\ 0 \le \zeta \le \chi}} \sum_{k \in \mathbb{Z}_+^n} \frac{|\hbar_\zeta^o \Delta_\chi a_k|}{\prod_{j: \chi_j = 0}(k_j + 1)}. \qquad (8.29)$$

Proof. Given the sequence from the theorem, we define the function $A(x)$ by (8.4). This function naturally satisfies the assumptions of Theorem 6.4. To see that (8.22) is satisfied, we apply to the integrals in it the Fourier-Hardy inequality (5.26) but for lower dimensions (or (1.61) in dimension one). We then are within the scope of the integrability of the Fourier transforms of Theorem 6.4. The bridge (8.21) allows us to conclude that (8.26)–(8.29) take place, in each of the appropriate cases, just following from (6.5)-(6.8). □

As above for functions, using similar embeddings for sequences, we can establish the following consequences of Theorem 8.25.

Corollary 8.30. *Let a be a sequence with bounded Hardy variation on $\mathbb{Z}_+^n$ and let a vanish at infinity along with all $\Delta^\eta a$ except $\eta = \mathbf{1}$. Then for $\eta \ne \mathbf{0}$, we have*

$$\int_{\mathbb{T}_+^n}\left|\sum_{k \in \mathbb{Z}_+^n} a_k \prod_{i:\eta_i=1} \cos x_i k_i \prod_{i:\eta_i=0} \sin x_i k_i\right| dx \lesssim \sum_{0 \le \chi \le 1-\eta} \sum_{k \in \mathbb{Z}_+^n} \left(\prod_{i:\chi_i=1} \frac{1}{k_i}\right)$$

$$\times \left[\left(\prod_{j:\chi_j=0} \frac{1}{k_j} \sum_{k_j \le m_j < 2k_j}\right) |\Delta_{\mathbf{1}-\chi} a_{m_{\mathbf{1}-\chi}, k_\chi}|^q\right]^{\frac{1}{q}}, \qquad (8.31)$$

and for $\eta = \mathbf{0}$, we have (8.28), with

$$\int_{\mathbb{T}_+^n} |F(x)|\, dx \lesssim \sum_{k \in \mathbb{Z}_+^n} \left[\left(\prod_{j=1}^n \frac{1}{k_j} \sum_{k_j \le m_j < 2k_j}\right) |\Delta_{\mathbf{1}} a_m|^q\right]^{\frac{1}{q}}, \qquad (8.32)$$

for some $1 < q < \infty$.

Corollary 8.33. *Let a be a sequence with bounded Hardy's variation on $\mathbb{Z}_+^n$ and let a vanish at infinity along with all $\Delta^\eta a$ except $\eta = \mathbf{1}$. Then for $\eta \ne \mathbf{0}$, we have*

$$\int_{\mathbb{R}_+^n}\left|\sum_{k \in \mathbb{Z}_+^n} a_k \prod_{i:\eta_i=1} \cos x_i k_i \prod_{i:\eta_i=0} \sin x_i k_i\right| dx$$

$$\lesssim \sum_{0 \le \chi \le 1-\eta} \sum_{k \in \mathbb{Z}_+^n} \left(\prod_{i:\chi_i=1} \frac{1}{k_i}\right) \max_{\substack{k_j \le m_j < 2k_j \\ j:\chi_j=0}} |\Delta_{\mathbf{1}-\chi} a_{m_{\mathbf{1}-\chi}, k_\chi}|, \qquad (8.34)$$

and for $\eta = \mathbf{0}$, we have (8.28), with

$$\int_{\mathbb{T}_+^n} |F(x)|\, dx \lesssim \sum_{k \in \mathbb{Z}_+^n} \max_{\substack{k_j \leq m_j < 2k_j \\ j=1,2,\ldots,n}} |\Delta_1 a_m|. \qquad (8.35)$$

Here the main feature is as follows. Till recently there existed certain multi-dimensional results for the Fourier transform (see, e.g., Chapter 6) of a function with bounded Hardy variation. Similarly, there existed multidimensional extensions for trigonometric series with coefficients of bounded Hardy type variation. However, there was no way to tie it as above in dimension one, since the existing bridges claimed an additional variation for the boundedness. Now, with the obtained temporary and stable bridges in hand, this became possible.

8.2 On the Poisson summation formula

There are specific versions of the Poisson summation formula for functions with bounded variation; see, e.g., [211, Ch.II, §13], [196, Lemma 2], or [131].

This section is devoted to even more special versions of the Poisson summation for functions of bounded variation. Our proof will be inductive and its one-dimensional version will be a part of the general proof, its basis. Also, contrary to the previous chapters, this will be related to Tonelli's variation rather than to Hardy's variation.

8.2.1 Background

In [211, Ch.II, §13], one can find certain versions of the Poisson summation formula for the validity of which the boundedness of the variation of a function is assumed along with additional conditions. In [196, Lemma 2], a somewhat different formula is obtained under the only assumption of bounded variation. With the Fourier transform defined as in (1.11), it reads as follows.

Lemma 8.36. *If f is a function of bounded variation on $\mathbb{R}$, written as $\|f\|_{BV} < \infty$,*

$$f(k) := \frac{1}{2}\left(f(k+0) + f(k-0)\right) \text{ for } k \in \mathbb{Z},$$

and $\lim_{|t| \to \infty} f(t) = 0$, *then for all* $x \not\equiv 0 (\mathrm{mod} 2\pi)$ *we have*

$$\sum_{k=-\infty}^{+\infty} f(k) e^{ikx} = \sum_{k=-\infty}^{+\infty} \widehat{f}(2k\pi - x). \qquad (8.37)$$

In [131], a multidimensional version of (8.37) is given under quite restrictive assumptions. But these were suitable for solving a problem in harmonic analysis. However, a question arises whether one can relax those conditions. Our goal is

8.2. On the Poisson summation formula 173

to find more or less minimal conditions to ensure a multidimensional extension of (8.37). It is worth mentioning that we study the problem only for one type of convergence of series and integrals; interesting phenomena for various types of convergence are discussed, e.g., in a recent paper [150]. Naturally, a problem of an appropriate notion of bounded variation arises here. In fact, our conditions in the main result are relaxed so that the class to which Lemma 8.36 is extended is wider than the known classes. The closest is the class of functions of bounded Tonelli variation; however, even in this case our class is wider and less restrictive.

8.2.2 A version of the Poisson summation formula

We prove the following multidimensional generalization of Lemma 8.36. Denote by $f(x \pm 0e_i)$, $i = 1, 2, \ldots, n$, the right (when $+0$) and left (when -0) limits in the i-th variable of f at the point x. More precisely,

$$f(x + 0e_i) = \lim_{t \to x_i + 0} f(x_1, \ldots, x_{i-1}, t, x_{i+1}, \ldots, x_n)$$

and

$$f(x - 0e_i) = \lim_{t \to x_i - 0} f(x_1, \ldots, x_{i-1}, t, x_{i+1}, \ldots, x_n).$$

Let $\pi_j e$ denote the j-th vector of type $(\pm 0e_1, \ldots, \pm 0e_n)$ of the $j = 1, 2, \ldots, 2^n$ different combinations of n signs $\pm$ in the value $f(x \pm 0e_1 \pm \cdots \pm 0e_n) = f(x + \pi_j e)$. Analogously to Lemma 8.36, suppose that

$$f(k) = \frac{1}{2^n} \sum_{\pi_j,\ j=1,2,\ldots,2^n} f(k + \pi_j e),$$

where we sum over all choices π_j of the right and left hand limits being possible at $k \in \mathbb{N}^n$. We consider functions f for which $f(k + \pi_j e)$ is the same regardless of the order of taking the repeated limit.

Let the Fourier transform of f be defined by (5.6).

Theorem 8.38. *Let f be of bounded variation in each variable, almost everywhere with respect to the rest of $n - 1$ variables, and $\lim_{|x| \to \infty} f(x) = 0$. Let also one of these variations be Lebesgue integrable with respect to the rest of $n - 1$ variables. Then we have for any x with $0 < |x_j| \leq \pi$, $j = 1, 2, \ldots, n$,*

$$\sum_{k \in \mathbb{Z}^n} f(k) e^{i \langle x, k \rangle} = \sum_{k \in \mathbb{Z}^n} \widehat{f}(2\pi k - x). \tag{8.39}$$

Proof. We will prove (8.39) by induction on dimension n. The case $n = 1$ can be found in [196, Lemma 2] or in [198, 4.1.4]. Since it is used as a basis of induction and in order to make the argument complete and self-contained, we reproduce this proof.

By Dirichlet's test, the series on the left is uniformly convergent on $[-\pi, \pi]$ outside any neighborhood of zero, and the integral in the definition of $\widehat{f}$ converges likewise on $\mathbb{R}$. Set
$$s(x) = \sum_{k=-\infty}^{+\infty} f(k)e^{ikx} - \widehat{f}(-x)$$
and
$$\sigma(x) = \sum_{k \neq 0} \widehat{f}(2k\pi - x).$$
We are now going to analyze the function $s(x) - \sigma(x)$. With the discontinuity at zero being removable, we will show that it is continuous on $[-\pi, \pi]$ and has zero Fourier coefficients. This will yield $s(x) = \sigma(x)$ for $0 < |x| \leq \pi$, and hence for all $x \not\equiv 0 \pmod{2\pi}$, since $s(x) - \sigma(x)$ is 2π-periodic. For the cut-off function f_n that coincides with f on $[-n, n]$ and vanishes off $[-n, n]$, we have $\|f_n\|_{BV} \leq \|f\|_{BV}$. Denoting
$$s_n(x) = \sum_{k=-n}^{n} f(k)e^{ikx} - \int_{-n}^{n} f(t)e^{ixt}dt,$$
we apply (8.5) to f_n to get
$$|s_n(x)| \leq 2\|f\|_{BV}$$
for any $n \in \mathbb{N}$. Now, for any $m \in \mathbb{Z}$, we calculate the m-th Fourier coefficient of s:
$$c_m(s(\cdot)) = \lim_{n \to \infty} \frac{1}{2\pi} \int_{-\pi}^{\pi} s_n(t)e^{-imt}dt$$
$$= f(m) - \frac{1}{2\pi} \lim_{n \to \infty} \int_{-\pi}^{\pi} e^{-imt} \int_{-n}^{n} f(u)e^{itu}du\,dt.$$
To estimate the limit, let us split $[-\pi, \pi]$ into two parts: $[-\delta, \delta]$ and $[-\pi, -\delta) \cup (\delta, \pi]$. After certain calculations we will let δ tend to zero. On $[-\delta, \delta]$, we have
$$\lim_{n \to \infty} \int_{-\delta}^{\delta} e^{-imt}dt \int_{-n}^{n} f(u)e^{iut}du = 2\int_{-\infty}^{\infty} f(u)\frac{\sin(u-m)\delta}{u-m}du.$$
Since the integral of $\frac{\sin t\delta}{t}$ over any interval is bounded by an absolute constant, the last integral converges uniformly in δ and hence its limit as $\delta \to 0$ is zero. The integral in the definition of $\widehat{f}$ converges uniformly outside of $[-\delta, \delta]$, which again gives a possibility to pass to the limit under the integral sign. By this,
$$c_m(s(\cdot)) = f(m) - \lim_{\delta \to 0} \frac{1}{2\pi} \int_{\delta \leq |x| \leq \pi} \widehat{f}(t)e^{imt}dt. \qquad (8.40)$$

Let us now proceed to $\sigma(x)$. Since it is independent of any correction of f on a set of measure zero, we can think of f as right continuous. Integrating by parts,

8.2. On the Poisson summation formula

we obtain

$$\sigma_n(x) = \sum_{1\leq |k|\leq n} \widehat{f}(2k\pi - x) = \sum_{1\leq |k|\leq n} \frac{-i}{2k\pi - x} \int_{-\infty}^{\infty} e^{-it(2k\pi - x)} df(t).$$

All the summands are continuous when $|x| \leq \pi$. Using the equality

$$\frac{1}{2k\pi - x} = \frac{1}{2k\pi} + \frac{x}{2k\pi(2k\pi - x)},$$

we split σ_n into two sums, whereas we split σ into two series, correspondingly. The second series converges uniformly on $[-\pi, \pi]$ in virtue of the Weierstrass test. The sequence of partial sums of the first series

$$\sum_{1\leq |k|\leq n} \frac{-i}{2k\pi} \int_{-\infty}^{\infty} e^{-it(2k\pi - x)} df(t) = -\int_{-\infty}^{\infty} e^{itx} \left(\sum_{k=1}^{n} \frac{\sin 2k\pi t}{k\pi} \right) df(t)$$

is represented as the sequence of Lebesgue integrals with respect to the measure generated by the function f of bounded variation. Since the sums in these integrals are uniformly bounded, by Lebesgue's dominated convergence theorem, this sequence converges to the continuous function

$$\int_{-\infty}^{\infty} e^{itx} \varphi_0(t) df(t),$$

with

$$\varphi_0(t) = t - [t] - \frac{1}{2}$$

for t is non-integer valued and $\varphi_0(k) = 0$ for $k \in \mathbb{Z}$. Thus, σ_n converges to σ on $[-\pi, \pi]$, with $\sigma \in C[-\pi, \pi]$, and

$$2\pi c_m(\sigma(\cdot)) = \lim_{n\to\infty} \sum_{1\leq |k|\leq n} \int_{-\pi}^{\pi} \widehat{f}(2k\pi - x) e^{-imx} dx$$

$$= \lim_{n\to\infty} \int_{\pi\leq |t|\leq \pi(2n+1)} \widehat{f}(t) e^{imt} dt. \qquad (8.41)$$

The needed equality $c_m(s(\cdot)) = c_m(\sigma(\cdot))$ follows now from the Fourier inversion for the function f of bounded variation combining (8.40) and (8.41) (see, e.g., [211, Vol.II, Ch.XVI, §2] or [198, 3.1.17]), which completes the proof of the one-dimensional result.

Proceeding to higher dimension, we assume, without loss of generality, that the variation that is Lebesgue integrable with respect to the remaining $n - 1$ variables is the one in the first variable. Integrating the representation of $\widehat{f}$ by parts in the first variable as the Stieltjes integral, we obtain for $x_1 \neq 0$

$$\left| \int_{\mathbb{R}^n} f(u) e^{-i\langle x, u\rangle} du \right| \leq \frac{1}{|x_1|} \int_{\mathbb{R}^{n-1}} \left[\int_{\mathbb{R}} |df(x_1, x_2, \ldots, x_n)| \right] dx_2 \cdots dx_n. \qquad (8.42)$$

The inner integral on the right is the total variation in the first variable, which is Lebesgue integrable with respect to the remaining $n-1$ variables. Therefore $\widehat{f}$ exists in the improper sense and converges uniformly on every compactum off the set $|x_1| < \delta$, with δ small enough. Partial sums of the corresponding series

$$\sum_{k \in \mathbb{Z}^n} f(k) e^{-i \langle x, k \rangle}$$

are Riemann integral sums for the partial integrals in (5.6) and thus converges in the same manner. Note that applying Dirichlet's test in one variable gives less.

We now proceed to the induction argument from n to $n+1$. Suppose that for any $k \in \mathbb{Z}^n$ the function $f(k, \cdot)$ is of bounded variation with respect to the $n+1$-st variable and apply the one-dimensional result to begin as follows:

$$\sum_{k \in \mathbb{Z}^{n+1}} f(k, k_{n+1}) e^{i \langle x, k \rangle} e^{x_{n+1} k_{n+1}}$$

$$= \sum_{k \in \mathbb{Z}^n} \left[\sum_{k_{n+1} \in \mathbb{Z}} f(k, k_{n+1}) e^{i x_{n+1} k_{n+1}} \right] e^{i \langle x, k \rangle}$$

$$= \sum_{k \in \mathbb{Z}^n} \left[\sum_{k_{n+1} \in \mathbb{Z}} \int_{\mathbb{R}} f(k, v) e^{iv(x_{n+1} - 2\pi k_{n+1})} \, dv \right] e^{i \langle x, k \rangle}.$$

Because of the mentioned uniform convergence, we can freely change the order of summation. Hence, the right-hand side is

$$\sum_{k_{n+1} \in \mathbb{Z}} \left[\sum_{k \in \mathbb{Z}^n} \int_{\mathbb{R}} f(k, v) e^{iv(x_{n+1} - 2\pi k_{n+1})} \, dv \, e^{i \langle x, k \rangle} \right]. \tag{8.43}$$

The function

$$F(u_1, \ldots, u_n, t) = \int_{\mathbb{R}} f(u_1, \ldots, u_n, v) e^{itv} \, dv$$

satisfies the assumptions of the theorem with respect to the first n variables. With this in hand, by the induction hypothesis we get for (8.43)

$$\sum_{k_{n+1}} \sum_{k \in \mathbb{Z}^n} \int_{\mathbb{R}^n} \int_{\mathbb{R}} f(u, v) e^{i \langle u, x - 2\pi k \rangle} e^{iv(x_{n+1} - 2\pi k_{n+1})} dv \, du = \sum_{k \in \mathbb{Z}^{n+1}} \widehat{f}(2\pi k - x),$$

the desired result. □

8.2.3 Concluding remarks and an example

Recall that a function f is of bounded Tonelli variation if it satisfies two types of conditions (see [187] or [33] or simply Subsection 5.2.2 in Chapter 5):

– almost everywhere in $(x_1, \ldots, x_{j-1}, x_{j+1}, \ldots, x_n)$, it is of bounded variation in one variable x_j for all $1 \leq j \leq n$,

8.2. On the Poisson summation formula

– these variations are Lebesgue integrable as functions of the other $n-1$ variables $x_1,\ldots,x_{j-1}, x_{j+1},\ldots,x_n$.

In our result, we continue to keep all the conditions of the first type, and it seems that there is no way to get rid of any part of them. As for the conditions of the second type, we assume only one of them from n. On the one hand, we cannot restrict ourselves to the conditions of the first type alone. On the other hand, what we have added to them seems to be really minimal.

Let us give an example of a function of two variables $f(x_1, x_2)$ such that it is not of bounded Tonelli variation but satisfies the assumptions of Theorem 8.38. This means that (8.39) cannot be derived from the corresponding result in [131] but nevertheless holds true in virtue of Theorem 8.38. More precisely, the first partial derivatives of f will be integrable in the same variable for (almost) all values of the other variable, but only one of these derivatives will be integrable over $\mathbb{R}^2$.

Therefore, let for $x_1, x_2 > 1$,

$$f(x_1, x_2) = \frac{\sin(x_1^\alpha x_2^\beta)}{x_1^p x_2^q},$$

with $p, q > 1$, $0 < p - \alpha \leq 1$ and $q - \beta > 1$, and zero otherwise. Then

$$f_{x_1}(x_1, x_2) = \alpha \frac{\cos(x_1^\alpha x_2^\beta)}{x_1^{p-\alpha+1} x_2^{q-\beta}} - p \frac{\sin(x_1^\alpha x_2^\beta)}{x_1^{p+1} x_2^q},$$

and

$$f_{x_2}(x_1, x_2) = \beta \frac{\cos(x_1^\alpha x_2^\beta)}{x_1^{p-\alpha} x_2^{q-\beta+1}} - q \frac{\sin(x_1^\alpha x_2^\beta)}{x_1^p x_2^{q+1}}.$$

The second summands on the right-hand sides of each formula are definitely integrable; therefore, only the first one should be analyzed. The choice of the parameters p, q, α, β ensures the claimed properties.

There is a different simple way to understand the spirit of this theorem. Let us take

$$f(x) = f_1(x_1) \cdots f_n(x_n),$$

with each f_j, $j = 1, 2, \ldots, n$, of bounded variation. For f to be of bounded Tonelli variation means that each f_j is also integrable, which means that f is integrable, not a natural assumption for a function of bounded variation. In our theorem, we assume in this case that one of them is not necessarily integrable, and, correspondingly, f is not necessarily integrable.

Chapter 9
Multidimensional case: radial functions

Of course, the approach elaborated in Part I plays a good part in the extension in the previous chapters of this second part. However, there is one more particular setting that can be treated immediately. Using a formula for the Fourier transform of a radial function from [139] (see also [92, Ch.4]), we can generalize the obtained above results to the radial case. Before doing this, we not only present certain needed preliminaries but also give a general necessary condition for the integrability of the Fourier transform. It is not for just radial functions, quite the contrary, it is also for general functions, but it is given in terms of the radial part of the given function. Also, a certain notion from Part I is used to provide terms for the formulation of a necessary condition. The latter also relates this to functions of bounded variation, though formally the obtained necessary condition does not claim for any assumption concerning variation. All is interrelated under the sun. . .

As usual, results for radial functions are a sort of interagent between the one-dimensional case and the general multivariate one.

9.1 Fractional derivative and MV classes

We first need to dwell upon a notion of fractional derivative. Our choice of one of them among several others (see, e.g., [165]) will be partially explained later on. For $0 < \delta < 1$ and a locally integrable function g on $(0, \infty)$, define the fractional (Weyl type) integral of order δ by

$$W_\omega^\delta g(t) = \begin{cases} \frac{1}{\Gamma(\delta)} \int_t^\omega g(r)(r-t)^{\delta-1} dr, & 0 < t < \omega, \\ 0, & t \geq \omega, \end{cases}$$

and, following Cossar [47], a fractional Weyl derivative of order α by

$$g^{(\alpha)}(t) = \lim_{\omega \to \infty} -\frac{d}{dt} W_\omega^{1-\alpha} g(t)$$

when $0 < \alpha < 1$ and

$$g^{(\alpha)}(t) = \frac{d^p}{dt^p} g^{(\delta)}(t)$$

when $\alpha = p + \delta$ with $p = 1, 2, \ldots$, and $0 < \delta < 1$. One of the reasons that just this type of fractional integral (and derivative) is chosen is that the Weyl integral of a function with compact support has, in turn, compact support, unlike the better known Riemann–Liouville integral

$$R_\alpha(f_0; t) = \frac{1}{\Gamma(\alpha)} \int_0^t f_0(r) \, (t-r)^{\alpha-1} dr.$$

All these notions may be found, for example, in [14, Ch.13] (see also [165]).

Let α^* be the greatest integer less than α. If α is fractional $\alpha^* = [\alpha]$, while for α integer $\alpha^* = \alpha - 1$. Consider the class $MV_{\alpha+1}^b$, with $\alpha > 0$ and $b \geq 0$, of $C(0, \infty)$-functions satisfying the following conditions

$$g, g', \ldots, g^{(\alpha^*)} \quad \text{are} \quad \text{LAC} \quad \text{on} \quad (0, \infty);$$

$$\lim_{t \to \infty} g(t), \quad \lim_{t \to \infty} t^{\alpha+b} g^{(\alpha)}(t) = 0;$$

and

$$\|g\|_{MV_{\alpha+1}^b} := \sup_{t>0} |t^b g(t)| + \int_0^\infty |d[(t^{\alpha+b} g^{(\alpha)}(t)]| < \infty.$$

These classes, especially for $\alpha = \frac{n-1}{2}$, generalize the notion of function with bounded variation to the radial case. This will be better seen later on in the "radial" generalizations of the results in Part I.

9.2 Necessary conditions

We denote by $f_0(t)$ the radial part of the function f, more precisely, the average of f over the sphere of radius $t > 0$ and center at the origin

$$f_0(t) = \frac{1}{\sigma_n} \int_{\mathbb{S}^{n-1}} f(tx) \, d\theta, \qquad (9.1)$$

where $\mathbb{S}^{n-1}$ is the unit sphere in the n-dimensional Euclidean space of area

$$\sigma_n = \frac{2\pi^{\frac{n}{2}}}{\Gamma(\frac{n}{2})},$$

9.2. Necessary conditions

or, in other words, the surface of the unit ball B^n of volume

$$\omega_n = \frac{2\pi^{\frac{n}{2}}}{n\Gamma(\frac{n}{2})}.$$

We are now in a position to formulate and prove the necessary conditions.

Theorem 9.2. *Let $f \in W_0(\mathbb{R}^n)$, that is,*

$$f(x) = \int_{\mathbb{R}^n} g(u)\, e^{i\langle x, u\rangle}\, du,$$

with $g \in L^1 \mathbb{R}^n$. Then the radial part f_0 satisfies the following conditions:

$$f_0 \in C^{\frac{n-1}{2}}(0, \infty); \tag{9.3}$$

$$\lim_{t \to \infty} t^s f_0^{(s)}(t) = 0, \quad 0 \leq s \leq \frac{n-1}{2}, \tag{9.4}$$

and

$$\lim_{t \to 0+} t^s f_0^{(s)}(t) = 0, \quad 0 < s \leq \frac{n-1}{2}; \tag{9.5}$$

where s is any number in the given interval. In addition, for any $t > 0$, the T-transform of $f_0^{(\frac{n-1}{2})}$ exists.

Proof. We rewrite the radial part as

$$f_0(t) = \frac{1}{\sigma_n} \int_{\mathbb{R}^n} g(u) \int_{\mathbb{S}^{n-1}} e^{it\langle x, u\rangle}\, d\theta\, du.$$

It is well known that (see, e.g., [176, Ch.4])

$$\int_{\mathbb{S}^{n-1}} e^{i\langle x, u\rangle}\, d\theta = (2\pi)^{\frac{n}{2}} (t|u|)^{1-\frac{n}{2}} J_{\frac{n}{2}-1}(t|u|),$$

where J_ν is the Bessel function of order ν that admits the following integral representation

$$J_\nu(t) = \frac{t^\nu}{2^\nu \sqrt{\pi}\,\Gamma(\nu + \frac{1}{2})} \int_{-1}^{1} (1 - u^2)^{\nu - \frac{1}{2}} e^{itu}\, du. \tag{9.6}$$

In particular,

$$J_0(t) = \frac{1}{\pi} \int_{-1}^{1} (1 - u^2)^{-\frac{1}{2}} e^{itu}\, du. \tag{9.7}$$

We shall make use of the following properties of the Bessel functions: the asymptotic formula (see, e.g., [14, §7.13.1(3)]), for $t \to \infty$,

$$J_\nu(t) = \sqrt{\frac{2}{\pi t}} \cos\left(t - \frac{\pi\nu}{2} - \frac{\pi}{4}\right) + O\left(t^{-\frac{3}{2}}\right) \tag{9.8}$$

and the formulas for derivatives

$$\frac{d}{dt}\left[t^{\pm\nu}J_\nu(t)\right] = \pm t^{\pm\nu}J_{\nu\mp 1}(t). \tag{9.9}$$

In fact, (9.8) holds for all $t > 0$ but is (asymptotically) meaningful for large t. Thus we have

$$f_0(t) = \Gamma\left(\frac{n}{2}\right) 2^{\frac{n}{2}-1} \int_{\mathbb{R}^n} g(u)\, j_{\frac{n}{2}-1}(t|u|)\, du, \tag{9.10}$$

where

$$j_\nu(t) = \frac{J_\nu(t)}{t^\nu}$$

is the normalized Bessel function. To find the fractional derivative of f_0, we have to know such a derivative for e^{iBt}. We cannot differentiate

$$\int_t^A (r-t)^{-\delta} e^{iBr}\, dr$$

immediately; first integration by parts should be fulfilled to get $(r-t)$ in a positive power. Routine calculations yield

$$-\frac{d}{dt}\int_t^A (r-t)^{-\delta} e^{iBr}\, dr = e^{iBA}(A-t)^{-\delta} - iB\int_t^A (r-t)^{-\delta} e^{iBr}\, dr,$$

and therefore

$$\frac{d}{dt} e^{iBt} = |B| e^{iBt} \int_0^\infty r^{-\delta} e^{iBr}\, dr$$

times a constant. Since the Fourier transform of $u^{-\delta}$ at B is $|B|^{\delta-1}$ times a constant (see, e.g., [15]), we obtain

$$(e^{iBt})^{(s)} = \gamma_s |B|^s e^{iBt},$$

which yields

$$j_\nu^{(s)}(t) = \gamma_{\nu,s} \int_{-1}^1 (1-v^2)^{\nu-\frac{1}{2}} |v|^s e^{itv}\, dv.$$

It follows from the definition of $j_{\frac{n}{2}-1}(t)$ and from (9.9) and (9.8) that for any non-negative s,

$$j_{\frac{n}{2}-1}^{(s)}(t) = O(t^{-\frac{n-1}{2}}). \tag{9.11}$$

Therefore, for each $s \leq \frac{n-1}{2}$ and $t \in (0,\infty)$ there exists the s-th derivative of f_0 equal to

$$f_0^{(s)}(t) = \gamma_{n,s} \int_{\mathbb{R}^n} g(u)\, |u|^s \int_{-1}^1 (1-v^2)^{\frac{n-3}{2}} v^s e^{it|u|v}\, dv\, du. \tag{9.12}$$

Because of the integrability of g and (9.11), we have (9.3).

9.2. Necessary conditions

If $s < \frac{n-1}{2}$, then (9.4) holds true by (9.11) and the Lebesgue dominated convergence theorem. For $s = \frac{n-1}{2}$ the proof of (9.4) is twofold. For n odd, we just integrate by parts $\frac{n-1}{2}$ times in the inner integral of (9.12) and use the Riemann-Lebesgue lemma. For n even, we first integrate by parts $\frac{n-2}{2}$ times in the same integral. The only summand in the resulting formula that is questionable is

$$\int_{\mathbb{R}^n} g(u)\,(t|u|)^{\frac{1}{2}} \int_{-1}^{1} (1-v^2)^{-\frac{1}{2}}|v|^\beta e^{it|u|v}\,dv\,du,$$

where $\beta > \frac{1}{2}$. We can rewrite it as

$$\int_{\mathbb{R}^n} g(u)\,(t|u|)^{\frac{1}{2}} \Bigg[\int_{-1}^{1} (1-v^2)^{-\frac{1}{2}} e^{it|u|v}\,dv$$
$$- \int_{-1}^{1} (1-v^2)^{-\frac{1}{2}}(1-v^\beta) e^{it|u|v}\,dv \Bigg] du. \qquad (9.13)$$

In the second integral in the square brackets, we can integrate by parts at least one more time, and all becomes clear. The rest is, by (9.7),

$$\int_{\mathbb{R}^n} g(u)\,(t|u|)^{\frac{1}{2}} J_0(t|u|)\,du$$

times a constant. Applying now (9.8), we are again in a position to use the Riemann-Lebesgue lemma which completes the proof of (9.4).

For $t \to 0$, the estimate (9.11) allows us to apply the Lebesgue dominated convergence theorem as above. Since the integral representation for $j^{(s)}_{\frac{n}{2}-1}(t)$ is always bounded, the presence of t in the positive power leads to (9.5), that is, for $0 < s \le \frac{n-1}{2}$.

Let us go to the last assertion of the theorem. We will explicitly give the proof of existence of the T-transform of $f_0^{(\frac{n-1}{2})}$, that is, the existence of the integral in (1.94) for $g = f_0^{(\frac{n-1}{2})}$, only for n even. For n odd the proof goes along the same lines but is even easier, since no fractional derivatives are involved.

For any $\delta \in (0, \frac{t}{2})$, we have

$$\int_\delta^{\frac{t}{2}} \frac{f_0^{(\frac{n-1}{2})}(t+z) - f_0^{(\frac{n-1}{2})}(t-z)}{z}\,dz$$
$$= \gamma_n \int_{\mathbb{R}^n} g(u) \Bigg[\int_{-1}^{1} |u|^{\frac{n-1}{2}}(1-v^2)^{\frac{n-3}{2}}|v|^{\frac{n-1}{2}}$$
$$\times \int_\delta^{\frac{t}{2}} \frac{e^{iv(t+z)|u|} - e^{iv(t-z)|u|}}{z}\,dz\,dv \Bigg] du.$$

To apply the Lebesgue dominated convergence theorem as $\delta \to 0+$, it suffices to check the boundedness in δ and u of the expression in the square brackets.

Denoting it by $\Phi_t(\delta, u)$ and

$$\varphi(v) = (1-v^2)^{\frac{n-3}{2}}|v|^{\frac{n-1}{2}},$$

we then integrate in v by parts $\frac{n}{2}-1$ times (recall that n is even). Since integrated terms vanish there, we obtain

$$\Phi_t(\delta, u) = \gamma_n |u|^{\frac{1}{2}} \int_{-1}^{1} \varphi^{(\frac{n}{2}-1)}(v) \int_{\delta}^{\frac{t}{2}} \left[\frac{e^{iv(t+z)|u|}}{(t+z)^{\frac{n}{2}-1}} - \frac{e^{iv(t-z)|u|}}{(t-z)^{\frac{n}{2}-1}}\right] \frac{dz}{z} dv.$$

Like in (9.13), the main problem is reduced to estimating

$$|u|^{\frac{1}{2}} \int_{-1}^{1} (1-v^2)^{-\frac{1}{2}} \int_{\delta}^{\frac{t}{2}} \left[\frac{e^{iv(t+z)|u|}}{(t+z)^{\frac{n}{2}-1}} - \frac{e^{iv(t-z)|u|}}{(t-z)^{\frac{n}{2}-1}}\right] \frac{dz}{z} dv.$$

Keeping in mind that

$$(t\pm z)^{1-\frac{n}{2}} - t^{1-\frac{n}{2}} = O(zt^{-\frac{n}{2}}),$$

as in [194, Th.3], and taking into account (9.7) and (9.8), we arrive at proving the boundedness in δ and u of the integral

$$|u|^{\frac{1}{2}} \int_{-1}^{1} (1-v^2)^{-\frac{1}{2}} \int_{\delta}^{\frac{t}{2}} \frac{e^{iv(t+z)|u|} - e^{iv(t-z)|u|}}{z} dz\, dv$$

$$= 2i|u|^{\frac{1}{2}} \int_{-1}^{1} (1-v^2)^{-\frac{1}{2}} e^{itv|u|} \int_{\delta v|u|}^{\frac{tv|u|}{2}} z^{-1} \sin z \, dz\, dv.$$

But it is shown in [194] that in this case $e^{it|u|v}$ is multiplied by a function from $\mathrm{Lip}\frac{1}{2}$ in the L^1 norm. The required boundedness is obvious then, which completes the proof. □

For a discussion on the necessary condition in dimension one, see (1.16) in Chapter 1 and discourses after this formula.

9.3 Radial extensions

Let us first describe, for the introduced class, the passage from the multidimensional Fourier transform of a radial function to the one-dimensional Fourier transform of a related function; it can be found in [139] or [92, Ch.4]. The most important partial case will give us a possibility to extend the obtained one-dimensional results to the radial case in a natural way. Let us denote

$$F_\alpha(t) = t^{\frac{n-1}{2}} f_0^{(\alpha)}(t) \quad \text{and} \quad F := F_{\frac{n+1}{2}}(t).$$

9.3. Radial extensions

Theorem 9.14. *Let $f_0 \in MV_{\alpha+1}^b$ with $0 < \alpha \le \frac{n-1}{2}$ and $b = \frac{n-1}{2} - \alpha$. Then there holds, for the radial extension $f(x) = f_0(|x|)$ of f_0,*

$$\widehat{f}(x) = 2^{\frac{n+1}{2}} \pi^{\frac{n-1}{2}} (-1)^{\alpha^*+1} \frac{1}{|x|^{\alpha+\frac{n-1}{2}}} \int_0^\infty F_\alpha(t) \cos\left(|x|t - \frac{\pi(n+2\alpha-1)}{4}\right) dt$$

$$+ \frac{C_\alpha}{|x|^{\alpha+\frac{n+1}{2}}} F_\alpha\left(\frac{\pi}{2|x|}\right)$$

$$+ O\left(\frac{1}{|x|^{\alpha+\frac{n+1}{2}}} \int_0^\infty \min\left\{\frac{2|x|t}{\pi}, \frac{\pi}{2|x|t}\right\} |dF_\alpha(t)|\right). \tag{9.15}$$

Here C_α is an absolute constant depending only on α. In particular, if $f_0 \in MV_{\frac{n+1}{2}}^0$, then

$$\widehat{f}(x) = 2^{\frac{n+1}{2}} \pi^{\frac{n-1}{2}} (-1)^{[\frac{n+3}{2}]} \frac{1}{|x|^{n-1}} \int_0^\infty F(t) \cos\left(|x|t - \frac{\pi n}{2}\right) dt$$

$$+ \frac{C_{\frac{n-1}{2}}}{|x|^n} F\left(\frac{\pi}{2|x|}\right) + O\left(\frac{G(|x|)}{|x|^{n-1}}\right), \tag{9.16}$$

where

$$\int_0^\infty |G(r)|\, dr \le \|f_0\|_{MV_{\frac{n+1}{2}}^0}.$$

It is (9.16) that gives us a possibility to extend one-dimensional results on the integrability of the Fourier transform of a function of bounded variation in full generality by posing the corresponding assumptions on F. This shows that $MV_{\frac{n+1}{2}}^0$ qualifies for the most natural radial extension of the class of functions with bounded variation. However, the situation is not that simple. Even an extension of Theorem 3.2 is of more specific form than its prototype, in the sense that a leading term appears.

Theorem 9.17. *Let $f_0 \in MV_{\frac{n+1}{2}}$; assume additionally F to be locally absolutely continuous on $(0, \infty)$. Then for $|x| > 0$,*

$$\widehat{f}(x) = \frac{C_n}{|x|^n} F\left(\frac{\pi}{2|x|}\right) + O\left(\frac{G(|x|)}{|x|^{n-1}}\right),$$

where $\int_0^\infty |G(r)|\, dr \le \|F'\|_Q$.

We can be more precise by observing that the last norm is in Q_e for n even and in Q_o for n odd. The theorem follows readily by applying Theorem 3.2 to the one-dimensional Fourier transform in (9.16). It is completely clear how to generalize to the radial case many other results given above.

However, we can extend more advanced Theorems 2.8 and 2.20 but not their refined versions from Chapter 3. In some form, Theorem 2.8 has already been extended to the radial case in [139] but of course not the much more recent Theorem

2.20. In any case, we now have interesting opportunities to play with assumptions according to whether the dimension is even or odd. Indeed, if n is even we have

$$\cos\bigl(|x|t - \frac{\pi n}{2}\bigr) = (-1)^{\frac{n}{2}} \cos |x|t,$$

while for n odd we have

$$\cos\bigl(|x|t - \frac{\pi n}{2}\bigr) = (-1)^{\frac{n-1}{2}} \sin |x|t.$$

In the next formulation, C_n and D_n are constants, depending only on n, that may be different in different occurrences. In each asymptotic relation, G denotes a remainder term; it also may be different in different occurrences.

Theorem 9.18. *Let $f_0 \in MV_{\frac{n+1}{2}}$ and F be locally absolutely continuous on $(0, \infty)$. Let $f(x) = f_0(|x|)$ be its radial extension.*
If n is EVEN and $F' \in H_e^1(\mathbb{R}_+)$, then, for $|x| > 0$,

$$\widehat{f}(x) = \frac{C_n}{|x|^n} F\left(\frac{\pi}{2|x|}\right) + \frac{D_n}{|x|^n} \int_0^{\frac{\pi}{2|x|}} F'(t) \ln \frac{2t|x|}{\pi}\, dt + O\left(\frac{G(|x|)}{|x|^{n-1}}\right),$$

where

$$\int_0^\infty |G(r)|\, dr \leq \|F'\|_{H_e^1(\mathbb{R}_+)}.$$

If n is ODD and $F' \in H_o^1(\mathbb{R}_+)$, then, for $|x| > 0$,

$$\widehat{f}(x) = \frac{C_n}{|x|^n} F\left(\frac{\pi}{2|x|}\right) + O\left(\frac{G(|x|)}{|x|^{n-1}}\right),$$

where

$$\int_0^\infty |G(r)|\, dr \leq \|F'\|_{H_o^1(\mathbb{R}_+)}.$$

As for Theorem 3.5, its extension might be of interest only in the case of odd dimensions.

Theorem 9.19. *Let $f_0 \in MV_{\frac{n+1}{2}}$ and F be locally absolutely continuous on $(0, \infty)$. Let $f(x) = f_0(|x|)$ be its radial extension. If n is ODD, then, for $|x| > 0$,*

$$\widehat{f}(x) = \frac{C_n}{|x|^n} F\left(\frac{\pi}{2|x|}\right) + \frac{D_n}{|x|^{n-1}} \mathcal{H}_o B_s F'(x) + O\left(\frac{G(|x|)}{|x|^{n-1}}\right),$$

where

$$\int_0^\infty |G(r)|\, dr \leq \|F'\|_{L^1(\mathbb{R}_+)}.$$

Remark 9.20. It is worth mentioning that posing conditions on $F(t)$ is not the same as posing conditions on $f_0^{(\frac{n+1}{2})}(t)$. This is discussed and established in [139] (see also [92, Ch.4]).

9.3. Radial extensions

It should be of interest to return to the estimates in [139] or in [92, Ch.4], where more technical details are given, in order to represent the remainder term in (9.16) via the balance operator (1.122). However, it might be much more difficult technical task than those while proving Theorems 4.15 and 4.20.

In addition to [139], there are two more approaches for the Fourier transform of a radial functions to be represented via a one-dimensional Fourier transform (without Bessel functions as in (5.12)): Leray's formula (see, e.g., [165, Ch.5, Lemma 25.1′] and a recent refinement of it in [129]) and the one from [160]. However, just (9.15) seems to be perfectly suitable for generalizations to the radial case of the known one-dimensional results for the Fourier transform of a function of bounded variation.

Afterword

I apologize for again returning to our book [92]. It ends with the words ... *we thank the reader for his/her patience and temporarily leave the world of exposition for the one of discovery. We hope to be back soon.* Each of the two authors of [92] has indeed returned to what was poetically called the "world of discovery". Publication of this book demonstrates that the author decided to share some of the fruits of the "world of discovery" with the readers by the world of exposition. In conclusion, let us overview what is included in this book and discuss what is not included and what could have been, e.g., possible applications and maybe future directions.

First, let us survey as a whole picture what has been done in the previous pages. Given a function f of bounded variation, one-dimensional, the task was to wade through its Fourier transform. Assume that we immediately made sure that f vanishes at infinity and is (locally) absolutely continuous. Integrating by parts, we readily obtained a test for checking whether the Fourier transform is integrable. Though we have certainly only performed a standard operation, surprisingly, the space of integrability was indicated really smaller than L^1 in which the derivative of f lives. However, assume that we cannot use the criterion for the integrability, or we need to apply this transform to a problem in approximation, harmonic analysis, or something else, where such a criterion, like many criteria in various problems, is too general to be handy. We instead check whether the derivative belongs to some Hardy space, or some space "below" the Hardy level, or "above". Then we have a clear measure of how far away this Fourier transform is from the integrability.

But let us assume that the above was only a part of the problem, and in fact the next step is similar but multi-dimensional. To stay within our scope, we first check whether the given function is of bounded Hardy variation. If a function is a product of one-dimensional functions, the problem hardly differs from the single-dimensional one. But we have established that in the arbitrary case, for many natural classes, the problem is still almost one-dimensional! More precisely, it reduces to the (operator) product of one-dimensional results, without any loss of sharpness.

Until recently, there was a path in the forest of the Fourier transforms of functions with bounded variation - a very broad path, with rich flora on both sides but with no vision of the whole forest. With the results of this text in hand, we do have a picture of the forest. Not that all the trees are cut down. On the one hand, we have more paths such that many points in the forest are reachable. On the other hand, we have a sort of vision from above. The forest, still having some mysteries and dark places, has become inhabited. So to say, no need to enter the forest from the very beginning when trying a new problem, one can have a base in a comfortable clearing in close vicinity of the problem.

Using geographical terms again, we can give a more global, "continental" picture. There is a continent of Fourier transforms in which countries of the "Fourier transform" civilization are located, the country of functions of bounded variation among them. In the "central" area of this country, the borders of which extend to the Hardy forests and amalgamated mountains, life is more or less well ordered and well organized. Hilbert transforms technologies make their life calm and comfortable. However, in this country beyond the mentioned borders, there are the lands that are not entirely terra incognita but are less studied and less explored, with very few travelers visiting them. Moreover, high (in dimension) in the mountains there are countries of the same civilizations but having more freedom and being more complicated. The country of functions of bounded Hardy variation is closer to the former and has more relations with them. Its inhabitants also use the Hilbert transform technologies, which makes the two countries not only well tied "economically" but also makes their languages very similar as well as other "cultural" aspects.

What is geography without maps? Ours may look as follows.

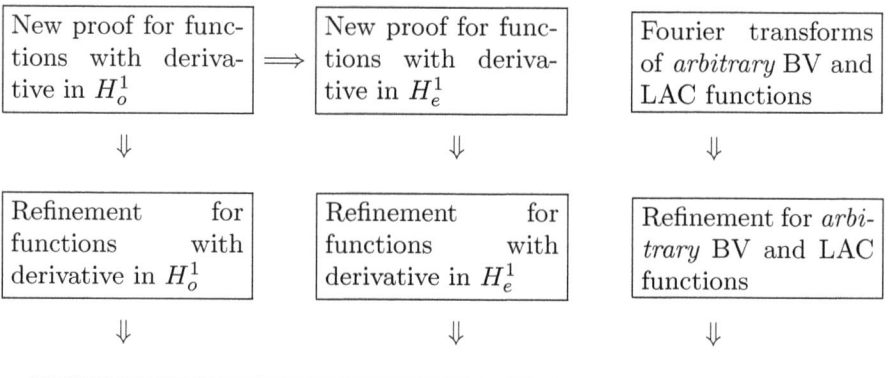

It seems to me that many surprises and discoveries await us in this continent. For example, in [4] affirmative answers are given to the open question of the second and third authors: *Is every function of bounded Hardy-Krause variation Borel measurable? Is the recently introduced $\mathcal{D}$-variation bounded for such a function?* Maybe for this new variation the problems considered in this book will become more colorful. Or, say, finding convenient spaces of integrability of the Fourier transform of a function of bounded variation with derivative far beyond any of Hardy spaces may bring bright prospects in both the problem itself and applications. As already mentioned, other variations than the Hardy one are almost not touched. For example, there are certain results related to Tonelli's variation and more that may definitely be obtained along the lines similar to the considered ones. However, an intention was not to load a story with unnecessary details. Let us mention two more open problems that seem to be highly interesting and important. In Chapter 3, a problem of intermediate spaces between two Hardy type spaces is posed and partially solved. However, it is still of importance to find

Afterword

more convenient intermediate spaces. In the same chapter, an asymptotic formula is found for the SINE Fourier transform of an arbitrary locally absolutely continuous function of bounded variation. It seems extremely interesting to find such a formula for the cosine Fourier transform. The result of Section 8.9 in Chapter 2 shows that there is a likelihood for existence of such a counterpart. Also, to generalize the Szökefalvi-Nagy type theorem in Section 7.3 to the case of more advanced singularities than point-wise is a challenging problem.

Let Vladimir Vysotsky finish this book on behalf of the author, partially with Alice, since he composed a musical on Alice's Adventures in Wonderland. Asking

Fairy tale, what remains of it
After the end, when it's told us completely?

he answers (in the same musical):

There is much unclear in that funny land.
It's so easy to get lost and tangled,

and then promises (in a song not from that musical)

I'll for sure come back,
Full of friends and of dreams.

Bibliography

[1] W. Abu-Shammala and A. Torchinsky, *The atomic decomposition in $L^1(\mathbf{R}^n)$*, Proc. Amer. Math. Soc. **135** (2007), 2839–2843.

[2] W. Abu-Shammala and A. Torchinsky, *Spaces between H^1 and L^1*, Proc. Amer. Math. Soc. **136** (2008), 1743–1748.

[3] C.R. Adams and J. A. Clarkson, *Properties of functions $f(x,y)$ of bounded variation*, Trans. Amer. Math. Soc. **36** (1934), 711–730.

[4] C. Aistleitner, F. Pausinger, A.M. Svane and R.F. Tichy, *On functions of bounded variation*, Math. Proc. Cambridge Philos. Soc. **162** (2017), 405–418.

[5] N.I. Akhiezer, *The classical moment problem and some related questions in analysis*, Oliver Boyd, 1965.

[6] G. Alexits, *Convergence problems of orthogonal series*, Akadémiai Kiadó, Budapest, 1961.

[7] L. Ambrosio, N. Fusco and D. Pallara, *Functions of Bounded Variation and Free Discontinuity Problems*, Clarendon Press, Oxford, 2000.

[8] K. Andersen, *Weighted Norm Inequalities for Hilbert Transforms and Conjugate Functions of Even and Odd Functions*, Proc. Amer. Math. Soc., Vol. 56, No. 1 (1976), 99–107.

[9] J. Appell, J. Banas, and N. J. Merentes Diáz, *Bounded Variation and Around*, De Gruyter series in nonlinear analysis and applications, De Gruyter, 2013.

[10] B. Aubertin and J.J.F. Fournier, *Integrability theorems for trigonometric series*, Studia Math. **107** (1993), 33–59.

[11] K.I. Babenko, *An inequality in the theory of Fourier integrals*, Izv. Akad. Nauk SSSR Ser. Mat. **25** (1961), 531–542 (Russian).

[12] S. Baron, E. Liflyand and U. Stadtmüller, *Complementary Spaces and Multipliers of Double Fourier Series for Functions of Bounded Variation*, J. Math. Anal. Appl. **250** (2000), 706–721.

[13] N.K. Bary, *A Treatise on Trigonometric Series, I and II*, MacMillan, New York, 1964.

[14] G. Bateman and A. Erdélyi, *Higher Transcendental Functions, Vol. II*, McGraw Hill Book Company, New York, 1953.

[15] H. Bateman and A. Erdélyi, *Tables of integral transforms, Vol. II*, McGraw Hill Book Company, New York, 1954.

[16] L. Bausov, *On linear methods for the summation of Fourier series*, Mat. Sbornik **68** (1965), 313–327 (Russian).

[17] W. Beckner, *Inequalities in Fourier analysis*, Ann. Math. **102** (1975), 159–182.

[18] E.S. Belinsky, *On asymptotic behavior of integral norms of trigonometric polynomials*, Metric Questions of the Theory of Functions and Mappings, Nauk. Dumka, Kiev **6** (1975), 15–24 (Russian).

[19] E.S. Belinsky, *Application of the Fourier transform to summability of Fourier series*, Sib. Mat. Zh. **XVIII** (1977), 497–511 (Russian). – English transl. in Siberian Math. J. **18** (1977), 353–363.

[20] E. Belinsky, E. Liflyand and R. Trigub, *The Banach algebra A^* and its properties*, J. Fourier Anal. Appl. **3** (1997), 103–129.

[21] J.J. Benedetto and G. Zimmermann, *Sampling multipliers and the Poisson summation formula*, J. Fourier Anal. Appl. **3** (1997), 505–523.

[22] E. Berkson and T.A. Gillespie, *Absolutely continuous functions of two variables and well-bounded operators*, J. London Math. Soc. (2), **30** (1984), 305-324.

[23] A. Beurling, *On the spectral synthesis of bounded functions*, Acta Math. **81** (1949), 225–238.

[24] R.P. Boas, *Absolute convergence and integrability of trigonometric series*, J. Rational Mech. Anal. **5** (1956), 621–632.

[25] R.P. Boas, *Quasi-positive sequences and trigonometric series*, Proc. London Math. Soc. **14A** (1965), 38–46.

[26] S. Bochner, *Summation of multiple Fourier series by spherical means*, Trans. Amer. Math. Soc. **40** (1936), 175–207 .

[27] S. Bochner, *Lectures on Fourier Integrals*, Princeton Univ. Press, Princeton, N. J., 1959.

[28] D. Borwein, *Linear functionals connected with strong Cesáro summability*, J. London Math. Soc. **40**(1965), 628–634.

[29] M. Bownik, *Boundedness of operators on Hardy spaces via atomic decompositions*, Proc. Amer. Math. Soc. **133** (2005), 3535–3542.

Bibliography

[30] S. Boza and M.J. Carro, *Discrete Hardy spaces*, Studia Math. **129** (1998), 31–50.

[31] J.S. Bradley, *Hardy inequalities with mixed norms*, Canad. Math. Bull. **21** (1978), 405–408.

[32] A. Brudnyi and Yu. Brudnyi, *On Banach structure of multivariate BV spaces I*, arXiv:1806.08824v1 [math.FA] 22 Jun 2018.

[33] Yu. Brudnyi, *Multivariate functions of bounded (k, p)-variation*, Banach Spaces and their Applications in Analysis, de Gruyter, 2007, 37–57.

[34] M. Buntinas and N. Tanović-Miller, *New integrability and L^1-convergence classes for even trigonometric series II*, Approximation Theory (J. Szabados and K Tandori, eds.), Colloq. Math. Soc. János Bolyai, North-Holland, Amsterdam **58** (1991), 103–125.

[35] P.L. Butzer, R.J. Nessel and W. Trebels, *On radial M_p^q Fourier multipliers*, Math. Struct., Comput. Math., Math. Modelling, Sofia, 1975, 187–193.

[36] P.L. Butzer and R.J. Nessel, *Fourier Analysis and Approximation. Volume 1. One-Dimensional Theory*, Academic Press, New York and London, 1971.

[37] A.P. Calderón and A. Zygmund, *On the differentiability of functions which are of bounded variation in Tonelli's sense*, Rev. Un. Mat. Arg. **20** (1962), 102–121.

[38] C. Carton-Lebrun and M. Fosset, *Moyennes et quotients de Taylor dans BMO*, Bull. Soc. Roy. Sci. Liege **53** (1984), 85–87.

[39] W. Cauer, *The Poisson integral for functions with positive real part*, Bull. Amer. Math. Soc. **38** (1932), 713–717.

[40] S.-Y. A. Chang and R. Fefferman, *Some recent developments in Fourier analysis and H^p-theory on product domains*, Bull. Amer. Math. Soc. (N.S.) **12** (1985), 1-43.

[41] K.-K. Chen *Summation of the Fourier series of orthogonal functions*, Science Press, Peking, 1957.

[42] J.A. Cima, A.L. Matheson and W.T. Ross, *The Cauchy transform*, Mathematical Surveys and Monographs, **125**, Amer. Math. Soc., Providence, RI, 2006.

[43] J.A. Clarkson, C.R. Adams, *On definitions of bounded variation for functions of two variables*, Trans. Amer. Math. Soc. **35** (1934), 824–854.

[44] A. Cohen, W. Dahmen, I. Daubechies, R. DeVore, *Harmonic analysis of the space BV*, Rev. Mat. Iberoamericana **19** (2003), 235–263.

[45] R. Coifman and G. Weiss, *Extensions of Hardy spaces and their use in analysis*, Bull. Amer. Math. Soc. **83** (1977), 569–645.

[46] A. Cordoba, *La formule sommatoire de Poisson*, C. R. Acad. Paris, Ser. I **306** (1988), 373–376.

[47] J. Cossar, *A theorem on Cesàro summability*, J. London Math. Soc. **16**(1941), 56–68.

[48] H. Davenport, *Note on irregularities of distribution*, Mathematika **1** (1954), 73–79.

[49] K.M. Davis and Y.-C. Chang, *Lectures on Bochner-Riesz means*, London Math. Soc. Lecture Note Series 114, Cambridge Univ. Press, Cambridge, 1987.

[50] J. Duoandikoetxea, *Fourier Analysis*, Graduate Studies in Mathematics, 29, Amer. Math. Soc., Providence, RI, 2001.

[51] R.E. Edwards, *Fourier series: a modern introduction*, Springer, Vol.1 – 1979, Vol.2 -1982.

[52] D. Faifman, *A characterization of Fourier transform by Poisson summation formula*, C. R. Acad. Paris, Ser. I **348** (2010), 407–410.

[53] R. Fefferman, *Some recent developments in Fourier analysis and H^P theory and product domains. II*, Function spaces and applications, Proc. US-Swed. Semin., Lund/Swed., Lect. Notes Math. **1302** (1988), 44–51.

[54] H.G. Feichtinger, *A characterization of Wiener's algebra on locally compact groups*, Arch. Math. (Basel), **29** (1977), 136–140.

[55] H.G. Feichtinger, *Wiener amalgams over Euclidean spaces and some of their applications*, Function spaces (Edwardsville, IL, 1990), Lect. Notes Pure Appl. Math., Dekker, New York **136**, 123–137 (1992).

[56] H.G. Feichtinger and F. Weisz, *Herz spaces and summability of Fourier transforms*, Math. Nachr. **281** (2008), 1–16.

[57] T.M. Flett, *Some theorems on odd and even functions*, Proc. London Math. Soc. (3) 8 (1958), 135-148.

[58] T.M. Flett, *Some elementary inequalities for integrals with applications to Fourier transforms*, Proc. London Math. Soc. (3), **29** (1974), 538–556.

[59] G.A. Fomin, *A Class of Trigonometric Series*, Matem. Zametki **23**(1978), 213–222 (Russian). – English transl. in Math. Notes **23** (1978), 117–123.

[60] J.J.F. Fournier and J. Stewart, *Amalgams of L^p and ℓ^q*, Bull. Amer. Math. Soc. **13**(1985), 1–21.

[61] S. Fridli, *Integrability and L^1-convergence of trigonometric and Walsh series*, Annales Univ. Sci. Budapest, Sect. Comp. **16** (1996), 149–172.

[62] S. Fridli, *Hardy spaces generated by an integrability condition*, J. Approx. Theory **113** (2001), 91–109.

Bibliography

[63] M. Ganzburg and E. Liflyand, *Estimates of best approximation and Fourier transforms in integral metrics*, J. Approx. Theory **83** (1995), 347–370.

[64] J. Garcia-Cuerva and J.L. Rubio de Francia, *Weighted Norm Inequalities and Related Topics*, North-Holland, 1985.

[65] J.B. Garnett, *Bounded Analytic Functions*, Springer, N.Y., 2007.

[66] D.V. Giang and F. Móricz, *On the integrability of trigonometric series*, Anal. Math. **18** (1992), 15–23.

[67] D.V. Giang and F. Móricz, *Lebesgue integrability of double Fourier transforms*, Acta Sci. Math. (Szeged) **58** (1993), 299–328.

[68] D.V. Giang and F. Móricz, *Multipliers of double Fourier transforms and Fourier series on L^1*, Acta Sci. Math. (Szeged) **58** (1993), 329–348.

[69] D.V. Giang and F. Móricz, *Multipliers of Fourier transforms and series on L^1*, Arch. Math. **62** (1994), 230–238.

[70] D.V. Giang and F. Móricz, *The Cesàro operator is bounded on the Hardy space H^1*, Acta Sci. Math. (Szeged) **61** (1995), 535–544.

[71] D.V. Giang and F. Móricz, *Hardy spaces on the plane and double Fourier transform*, J. Fourier Anal. Appl. **2** (1996), 487–505.

[72] E. Giusti, *Minimal surfaces and functions of bounded variation*, Birkhäuser, Boston, 1984.

[73] M.L. Goldman, *Estimates for Multiple Fourier transforms of radially symmetric monotone functions*, Sib. Mat. Zh. **18**(1977), 549–569 (Russian). – English transl. in Sib. Math. J. **18** (1977), 391–406.

[74] L. Grafakos, Classical Fourier Analysis, 2nd ed., Springer, 2008.

[75] L. Grafakos, Modern Fourier Analysis, 2nd ed., Springer, 2009.

[76] R.L. Graham, D.E. Knuth and O. Patashnik, *Concrete Mathematics*, AddisonWesley, Reading MA, 1988.

[77] K. Gröchenig, *Foundations of Time-Frequency Analysis*, Appl. Numer. Harmon. Anal. Birkhäuser, Boston, MA, 2001.

[78] Y. Han, G. Lu and K. Zhao, *Discrete Calderón's identity, atomic decomposition and boundedness criterion of operators on multiparameter Hardy spaces*, J. Geom. Anal. **20** (2010), 670–689.

[79] G.H. Hardy, *On double Fourier series, and especially those which represent the double zeta-function with real and incommensurable parameters*, Quart. J. Math. **37** (1906), 53–79.

[80] G.H. Hardy, *Oscillating Dirichlet's integrals*, Quart. J. Pure Appl. Math. **44** (1913), 1–40, 242–263.

[81] G.H. Hardy and J.E. Littlewood, *Some new properties of Fourier constants*, Math. Ann. **97** (1927), 159–209.

[82] G.H. Hardy and J.E. Littlewood, *Some more theorems concerning Fourier series and Fourier power series*, Duke Math. J. **2** (1936), 354–382.

[83] C. Heil, *An introduction to weighted Wiener amalgams*, Wavelets and their Applications (Chennai, 2002), M. Krishna, R. Radha and S. Thangavelu, eds., Allied Publishers, New Delhi, 183–216 (2003).

[84] H. Helson, *On a theorem of F. and M. Riesz*, Colloq. Math. **3** (1955), 113–117.

[85] H. Helson, *Harmonic Analysis*, Addison-Wesley Publ. Co., 1983.

[86] C.S. Herz, *Lipschitz spaces and Bernstein's theorem on absolutely convergent Fourier transforms*, J. Math. Mech. **18** (1968), 283–323.

[87] E. Hewitt and K. Stromberg, *Real and abstract analysis*, Springer-Verlag, Heidelberg/Berlin, 1965.

[88] E. Hille and J.D. Tamarkin, *On the absolute integrability of Fourier transforms*, Fund. Math. **25** (1935), 329–352.

[89] E.W. Hobson, *The theory of functions of a real variable and the theory of Fourier's series. Vol. 1*, third edition, University Press, Cambridge, 1927; Dover, New York, 1957.

[90] F. Holland, *Harmonic analysis on amalgams of L^p and l^q*, J. London Math. Soc. (2)**10** (1975), 295–305.

[91] K. Hoffman, *Banach spaces of analytic functions*, Prentice-Hall, Inc., Englewood Cliffs, N.J., 1962.

[92] A. Iosevich and E. Liflyand, *Decay of the Fourier transform: analytic and geometric aspects*, Birkhäuser, 2014.

[93] O.S. Ivashev-Musatov, *On Fourier-Stieltjes coefficients of singular functions*, Izv. Akad. Nauk SSSR, Ser.Mat. **20** (1956), 179–196 (Russian). – English transl. in AMS, 1958, 18 p.

[94] S.I. Izumi and T. Tsuchikura, *Absolute convergence of trigonometric expansions*, Tôhoku Math. J. **7** (1955), 243–251.

[95] B. Jawerth and A. Torchinsky, *A note on real interpolation of Hardy spaces in the polydisk*, Proc. Amer. Math. Soc. **96** (1986), 227–232.

[96] R.L. Johnson and C.R. Warner, *The convolution algebra $H^1(R)$*, J. Function Spaces Appl. **8** (2010), 167–179.

[97] J.-P. Kahane, *Séries de Fourier absolument convergentes*, Springer-Verlag, 1970.

[98] J.-P. Kahane, *Stylianos Pichorides*, J. Geom. Anal. **3** (1993), 533–542.

[99] Y. Katznelson, *An Introduction to Harmonic Analysis. 2nd corr. ed.*, Dover Publications Inc, New York, 1976.

[100] F.W. King, *Hilbert transforms, Vol.1*, Enc. Math/ Appl., Cambridge Univ. Press, Cambridge, 2009.

[101] H. Kober, *A note on Hilberts operator*, Bull. Amer. Math. Soc., **48** (1942), 421–426.

[102] H. Kober, *A note on Hilbert transforms*, Quart. J. Math., Oxford Ser. **14** (1943), 49–54.

[103] A.N. Kolmogorov, *Une série de Fourier-Lebesgue divergente partout*, C. R. Acad. Sci. Paris **183** (1926), 1327–1328.

[104] P. Koosis, *Proof of a theorem of the brothers Riesz*, Studia Math. **17** (1958), 295–298.

[105] P. Koosis, *Introduction to H_p Spaces*, Cambridge Univ. Press, 1980.

[106] M.G. Krein and A.A. Nudelman, *The Markov moment problem and extremal problems. Ideas and problems of P.L. Chebyshev and A.A. Markov and their further development*, Translations of Mathematical Monographs, Vol. 50, American Mathematical Society, Providence, R.I., 1977.

[107] E. Liflyand, *Fourier transform of functions from certain classes*, Anal. Math. **19** (1993), 151–168.

[108] E. Liflyand, *Fourier transforms of radial functions*, Integral Transforms and Special Functions **4** (1996), 279–300.

[109] E. Liflyand, *A family of function spaces and multipliers*, Israel Math. Conf. Proc. **13** (1999), 141–149.

[110] E. Liflyand, *On Bausov-Telyakovskii type conditions*, Georgian Math. J. **47** (2000), 753–764.

[111] E. Liflyand, *On quasi-monotone functions and sequences*, Comput. Methods Funct. Theory **1** (2001), 345–352.

[112] E. Liflyand, *Lebesgue constants of multiple Fourier series*, Online J. Anal. Combin. **1** (2006), 112 p.

[113] E. Liflyand, *A counterexample on embedding of spaces*, Georgian Math. J., **16** (2009), 757–760.

[114] E. Liflyand, *Fourier transform versus Hilbert transform*, Ukr. Math. Bull. **9** (2012), 209–218. Also published in J. Math. Sciences **187** (2012), 49–56.

[115] E. Liflyand, *Fourier transforms on an amalgam type space*, Monatsh. Math. **172** (2013), 345–355.

[116] E. Liflyand, *Hausdorff operators on Hardy spaces*, Eurasian Math. J. **4** (2013), 101–141.

[117] E. Liflyand, *On Fourier re-expansions*, J. Fourier Anal. Appl. **20** (2014), 934–946.

[118] E. Liflyand, *Interaction between the Fourier transform and the Hilbert transform*, Acta Comm. Uni. Tartu. Math. **18** (2014), 19–32.

[119] E. Liflyand, *Variations on the theorems of F. and M. Riesz and of Hardy and Littlewood*, Georgian Math. J. **21** (2014), 337–341.

[120] E. Liflyand, *On an Sz.-Nagy theorem*, Acta Math. Hungar. **145** (2015), 252–262.

[121] E. Liflyand, *Integrability spaces for the Fourier transform of a function of bounded variation*, J. Math. Anal. Appl. **436** (2016), 1082–1101.

[122] E. Liflyand, *Multiple Fourier transforms and trigonometric series in line with Hardys variation*, Contemporary Math. **659** (2016), 135–155.

[123] E. Liflyand, *Asymptotics of the sine Fourier transform of a function of bounded variation*, Mat. Zametki **100** (2016), 108-116 (Russian). – English transl. in Math. Notes **100** (2016), 93–99.

[124] E. Liflyand, *Asymptotic behavior of the Fourier transform of a function of bounded variation*, Novel methods in harmonic analysis with applications to numerical analysis and data processing, Applied and Numerical Harmonic Analysis series (ANHA), Birkhäuser/Springer, 2016.

[125] E. Liflyand, *Asymptotic behavior of the Fourier transform of a function with derivative in a Hardy space*, Pure Appl. Funct. Anal. **2** (2017), 117–128.

[126] E. Liflyand, *The Fourier transform of a function of bounded variation: symmetry and asymmetry*, J. Fourier Anal. Appl. **24** (2018), 525–544.

[127] E. Liflyand, *The Fourier transform of a convex function revisited*, NEAM, Theta, 2018, 81–91.

[128] E. Liflyand, S. Samko and R. Trigub, *The Wiener algebra of absolutely convergent Fourier integrals: an overview*, Anal. Math. Physics **2** (2012), 1–68.

[129] E. Liflyand and S. Samko, *On Leray's Formula*, Methods of Fourier Analysis and Approximation Theory, M. Ruzhansky and S. Tikhonov eds., Springer, 2016, 139–146.

[130] E. Liflyand, U. Stadtmüller and R. Trigub, *An interplay of multidimensional variations in Fourier Analysis*, J. Fourier Anal. Appl. **17** (2011), 226–239.

[131] E. Liflyand and U. Stadtmüller, *A multidimensional Euler-Maclaurin formula and application*, Proceedings of an international conference Complex Analysis and Dynamical Systems V, Israel Mathematical Conference Proceedings sub-series of Contemporary Mathematics, 2013, 181–191.

[132] E. Liflyand and U. Stadtmüller, *On a Hardy-Littlewood theorem*, Bull. Inst. Math. Acad. Sinica (New Series) **8** (2013), 481–489.

[133] E. Liflyand and U. Stadtmüller, *Poisson summation for functions of bounded variation on $\mathbb{R}^d$*, Acta Sci. Math. (Szeged) **80** (2014), 491–498.

[134] E. Liflyand and U. Stadtmüller, *A multidimensional Hardy-Littlewood theorem*, Topics in Classic and Modern Analysis. In memory of Yingkang Hu, Eds. M. Abell, E. Iacob, A. Stokolos, S. Taylor, S. Tikhonov and Jiehua Zhu, 2018.

[135] E. Liflyand and S. Tikhonov, *The Fourier transforms of general monotone functions*, Analysis and Mathematical Physics, Trends in Mathematics, Birkhäuser, 2009, 373–391.

[136] E. Liflyand and S. Tikhonov, *Weighted Paley-Wiener theorem on the Hilbert transform*, C.R. Acad. Sci. Paris, Ser. I **348** (2010), 1253–1258.

[137] E. Liflyand and S. Tikhonov, *A concept of general monotonicity and applications*, Math. Nachrichten **284** (2011), 1083–1098.

[138] E. Liflyand, S. Tikhonov and M. Zeltser, *Extending tests for convergence of number series*, J. Math. Anal. Appl. **377** (2011) 194–206.

[139] E. Liflyand and W. Trebels, *On asymptotics for a class of radial Fourier transforms*, Z. Anal. Anwendungen (J. Anal. Appl.) **17** (1998), 103–114.

[140] L.H. Loomis, *A note on Hilbert's transform*, Bull. Amer. Math. Soc. **52** (1946), 1082–1086.

[141] E. Lukacs, *Characteristic functions, 2nd ed.*. Charles Griffin & Co. Ltd., London, 1970.

[142] B. Makarov and A. Podkorytov, *Real Analysis: Measures, Integrals and Applications*, Springer, 2013.

[143] T. Mansour and M. Schork, *Commutation Relations, Normal Ordering, and Stirling Numbers*, CRC Press, Boca Raton, FL, 2015.

[144] J. Mashreghi, *Representation theorems in Hardy spaces*, LMS, Student Texts **74**, Cambridge Univ. Press, Cambridge, 2009.

[145] V. Matsaev and M. Sodin, *Distribution of Hilbert transforms of measures*, Geometric Funct. Anal., **10** (2000), 160–184.

[146] Y. Meyer, *Wavelets and operators*, Cambridge Univ. Press, 1992.

[147] A. Miyachi, *On some Fourier multipliers for $H^p(R^n)$*, J. Fac. Sci. Univ. Tokyo Sec. IA **27** (1980), 157–179.

[148] F. Móricz, *On the integrability and L^1-convergence of complex trigonometric series*, Proc. Amer. Math. Soc. **113** (1991), 53–64.

[149] F. Móricz, *On the integrability and L^l-convergence of double trigonometric series*, Studia Math. **98** (1991), 203–225.

[150] F. Móricz, *On the regular convergence of multiple integrals of locally Lebesgue integrable functions over $\overline{\mathbb{R}}_+^m$*, C.R. Math. Acad. Sci. Paris **350** (2012), 459–464.

[151] I.P. Natanson, *Theory of functions of a real variable*, Frederick Ungar Publishing Co., New York, 1955.

[152] F Natterer, *The mathematics of computerized tomography*, Classics in Applied Mathematics **32**, SIAM, Philadelphia, 2001.

[153] R. Nevanlinna, *Asymptotische Entwicklungen beschränkter Funktionen und das Stieltjessche Momentenproblem*, Ann. Acad. Sci. Fenn. Ser. A **18** (1922), 1–53.

[154] C.P. Niculescu and J.-E. Persson, *Convex Functions and Their Applications: A Contemporary Approach*, Springer, New York, 2006.

[155] R.E.A.C. Paley and N. Wiener, *Notes on the theory and application of Fourier transform*, Note II, Trans. Amer. Math. Soc. **35** (1933), 354–355.

[156] J.N. Pandey, *The Hilbert transform of Schwartz distributions*, Proc. Amer. Math. Soc. **89** (1983), 86–90.

[157] J.N. Pandey, *The Hilbert transform of Schwartz distributions and applications*, John Wiley & Sons, New York, 1996.

[158] M.C. Pereyra and L.A. Ward, *Harmonic analysis: from Fourier to wavelets*, Student Mathematical Library, 63. IAS/Park City Mathematical Subseries. Amer. Math. Soc., Providence, RI; Institute for Advanced Study (IAS), Princeton, NJ, 2012.

[159] A. Plessner, *Eine Kennzeichnung der totalstetigen Funktionen*, J. Reine Angew. Math. **160** (1929), 26–32.

[160] A.N. Podkorytov, *Linear means of spherical Fourier sums*, Operator Theory and Function Theory, ed. M. Z. Solomyak, Leningrad Univ., Leningrad **1** (1983), 171–177 (Russian).

[161] A. Poltoratski, *A problem on completeness of exponentials*, Ann. of Math. (2) **178** (2013), 983–1016.

[162] H. Reiter and J.D. Stegeman, *Classical harmonic analysis and locally compact groups. Second edition*, London Math. Soc. Monographs. New Series, —bf 22, The Clarendon Press, Oxford University Press, New York, 2000.

[163] W. Rudin, *Fourier analysis on groups*, Interscience Tracts in Pure and Applied Mathematics, No. 12 Interscience Publishers (a division of John Wiley and Sons), New York-London, 1962.

[164] D. Ryabogin and B. Rubin, *Singular integral operators generated by wavelet transforms*, Integr. Equ. Oper. Theory **35** (1999), 105–117.

[165] S.G. Samko, A.A. Kilbas, O.I. Marichev, *Fractional Integrals and Derivatives. Theory and Applications*, Gordon & Breach Sci. Publ., New York, 1992.

[166] G.E. Shilov, *On the Fourier coefficients of a class of continuous functions*, Doklady AN SSSR **35** (1942), 3–7 (Russian).

[167] G.E. Shilov, *Mathematical Analysis. A Special Course*, Pergamon Press Ltd., 1965.

[168] W.T. Sledd and D.A. Stegenga, *An H^1 multiplier theorem*, Ark. Mat. **19** (1981), 265–270.

[169] C.D. Sogge, *Fourier Integrals in Classical Analysis*, Cambridge Univ. Press, 1993.

[170] E.M. Stein, *On certain exponential sums arising in multiple Fourier series*, Ann. Math. **73** (1961), 87–109.

[171] E.M. Stein, *Note on the class $L\log L$*, Studia Math. **32** (1969), 305–310.

[172] E.M. Stein, *Singular Integrals and Differentiability Properties of Functions*, Princeton Univ. Presss, Princeton, N. J., 1970.

[173] E.M. Stein, *Harmonic Analysis: Real-Variable Methods, Orthogonality, and Oscillatory Integrals*, Princeton Univ. Presss, Princeton, N.J., 1993.

[174] E.M. Stein and R. Shakarchi, *Complex analysis*, Princeton Lectures in Analysis, II. Princeton University Press, Princeton, N.J., 2003.

[175] E.M. Stein and R. Shakarchi, *Functional Analysis: Introduction to Further Topics in Analysis*, Princeton Univ. Press, Princeton and Oxford, 2011.

[176] E.M. Stein and G. Weiss, *Introduction to Fourier analysis on Euclidean spaces*, Princeton Univ. Press, Princeton, N. J., 1971.

[177] C. Sweezy, *Subspaces of $L^1(\mathbb{R}^d)$*, Proc. Amer. Math. Soc. **132** (2004), 3599–3606.

[178] B. Sz-Nagy, *Sur une classe générale de procédés de sommation pour les séries de Fourier*, Hungarica Acth Math. **1** (1948), 14–52.

[179] M. Taibleson, *Fourier coefficients of functions of bounded variation*, Proc. Amer. Math. Soc. **18** (1967), 766.

[180] A.A. Talalyan and G.G. Gevorkyan, *Representation of absolutely continuous functions of several variables*, Acta Sci. Math. (Szeged) **54** (1990), 277-283 (Russian).

[181] S.A. Telyakovskii, *Integrability conditions for trigonometric series and their applications to the study of linear summation methods of Fourier series*, Izv. Akad. Nauk SSSR, Ser. Matem. **28** (1964), 1209–1236 (Russian).

[182] S.A. Telyakovskii, *An asymptotic estimate for the integral of the absolute value of a function, given by the sine series*, Sibirsk. Mat. Zh. **8** (1967), 1416–1422 (Russian).

[183] S.A. Telyakovskii, *An estimate, useful in problems of approximation theory, of the norm of a function by means of its Fourier coefficients*, Trudy Mat. Inst. Steklova **109** (1971), 65–97 (Russian). – English transl. in Proc. Steklov Math. Inst. **109** (1971), 73–109.

[184] S.A. Telyakovskii, *Concerning a sufficient condition of Sidon for the integrability of trigonometric series*, Mat. Zametki **14** (1973), 317–328 (Russian). – English transl. in Math. Notes **14** (1973), 742–748.

[185] S. Tikhonov, *Trigonometric series with general monotone coefficients*, J. Math. Anal. Appl. **326** (2007), 721–735.

[186] E.C. Titchmarsh, *Introduction to the theory of Fourier integrals*, Oxford, 1937.

[187] L. Tonelli, *Series trigonometriche*, Bologna, 1928 (Italian).

[188] W. Trebels, *Multipliers for (C,α)-Bounded Fourier Expansions in Banach Spaces and Approximation Theory*, Lect. Notes Math. 329, Springer, Berlin, 1973.

[189] W. Trebels, *Some Fourier multiplier criteria and the spherical Bochner-Riesz kernel*, Rev. Roumaine Math. Pures Appl. **20** (1975), 1173–1185.

[190] R.M. Trigub, *Several remarks on Fourier series*, (Russian) Mat. Zametki **3** (1968), 597–603 (Russian). – English transl. in Math. Notes **3** (1968), 380–383.

[191] R.M. Trigub, *On integral norms of polynomials*, Matem. Sbornik **101(143)** (1976), 315–333(Russian). – English transl. in Math. USSR Sbornik **30** (1976), 279–295.

[192] R.M. Trigub, *Integrability and asymptotic behavior of the Fourier transform of a radial function*, Metric Questions of Theory of Functions and Mappings, Kiev, Nauk. dumka, 1977, 142–163 (Russian).

[193] R.M. Trigub, *Linear methods of summability of simple and multiple Fourier series and their approximate properties*, Proc. Intern. Conf. Nauka, Moscow, 1977, 383-390 (Russian).

[194] R.M. Trigub, *Absolute convergence of Fourier integrals, summability of Fourier series, and polynomial approximation of functions on the torus*, Izv. Akad. Nauk SSSR, Ser.Mat. **44** (1980), 1378–1408 (Russian). – English translation in Math. USSR Izv. **17** (1981), 567–593.

[195] R.M. Trigub, *Multipliers of Fourier series and approximation of functions by polynomials in spaces C and L*, Dokl. Akad. Nauk SSSR **306** (1989), 292–296 (Russian). – English transl. in Soviet Math. Dokl. **39** (1989), 494–498.

[196] R.M. Trigub, *A Generalization of the Euler-Maclaurin formula*, Mat. Zametki **61** (1997), 312–316 (Russian). – English transl. in Math. Notes **61** (1997), 253–257.

[197] R.M. Trigub, *The Fourier transform of a quasi-convex function and that from V^**, Ukr. Math. Bull. **11** (2014), 274–286 (Russian). – English transl. in J. Math. Sciences **204** (2014), 369–378.

[198] R.M. Trigub and E.S. Belinsky, *Fourier Analysis and Appoximation of Functions*, Kluwer, 2004.

[199] F. Weisz, *Herz spaces and restricted summability of Fourier transforms and Fourier series*, J. Math. Anal. Appl. **344** (2008), 42–54.

[200] R. Wheeden and A. Zygmund, *Measure and Integral*, Pure and Applied Mathematics **49**, Marcel Dekker, Inc. N. Y. – Basel, 1977.

[201] D.V. Widder, *The Laplace transform*, Princeton, 1946.

[202] C.L. Williams, B. G. Bodmann and D. J. Kouri, *Fourier and beyond: invariance properties of a family of integral transforms*, arXiv:1403.4168v4 [math.FA] 14 Jul 2016.

[203] N. Wiener, *On the representation of functions by trigonometric integrals*, Math. Z. **24** (1926), 575–616.

[204] N. Wiener, *The Fourier integral and certain of its applications*, Dover Publ., Inc., New York, 1932.

[205] N. Wiener and A. Wintner, *Fourier-Stieltjes transforms and singular infinite convolutions*, Amer. J. Math. **60** (1938), 513–522.

[206] T. Wolff, *Math 191 lecture notes*, Harmonic analysis lecture notes; CalTech, 2000.

[207] J. Xiao, *L^p and BMO bounds of weighted Hardy-Littlewood averages*, J. Math. Anal. Appl. **262** (2001), 660–666.

[208] W.H. Young and G.C. Young, *On the discontinuities of monotone functions of several variables*, Proc. London Math. Soc., (2) **22** (1923), 124–142.

[209] W. Ziemer, *Weakly differentiable functions*, Springer Verlag, New York, 1989.

[210] A. Zygmund, *Some points in the theory of trigonometric and power series*, Trans. Amer. Math. Soc. **36** (1934), 586–617.

[211] A. Zygmund, *Trigonometric series, Vol. I, II*, Cambridge Univ. Press, Cambridge, U.K., 1966.

Basic notations

We present the list of basic notations in the order in which they appear in the text of the book. Some notation is not listed if it is introduced and used "locally", that is, in only a specific certain section or subsection.

$Vf := V_{[a,b]}f$ total variation of f on $[a, b]$

$f \in BV([a,b])$ f is of bounded variation

BV_0 functions of bounded variation vanishing at infinity

$AC([a,b])$ absolutely continuous functions on $[a, b]$

$LAC([a,b])$ locally absolutely continuous functions on $[a, b]$

$\widehat{f_c}(x) = \int_0^\infty f(t) \cos xt\, dt$ cosine Fourier transform

$\widehat{f_s}(x) = \int_0^\infty f(t) \sin xt\, dt$ sine Fourier transform

$W_0(\mathbb{R}) = \{f \in C_0(\mathbb{R}) : f(t) = \int_\mathbb{R} g(x) e^{itx} dx, g \in L^1(\mathbb{R})\}$ Wiener algebra

$\mathcal{H}g(x) = \frac{1}{\pi} \int_\mathbb{R} \frac{g(t)}{x-t} dt$ Hilbert transform

$\mathcal{H}_\delta g(x) = \frac{1}{\pi} \int_{|t-x|>\delta} \frac{g(t)}{x-t} dt$ truncated Hilbert transform

$\mathcal{H}_o g(x) = \frac{2}{\pi} \int_0^\infty \frac{tg(t)}{x^2-t^2} dt$ odd Hilbert transform

$\mathcal{H}_e g(x) = \frac{2}{\pi} \int_0^\infty \frac{xg(t)}{x^2-t^2} dt$ even Hilbert transform

$\lesssim$ and $\gtrsim$ abbreviations for $\leq C$ and $\geq C$, where $C > 0$ is a constant

$\hbar a(m) = \sum\limits_{\substack{k=-\infty \\ k \neq m}}^{\infty} \frac{a_k}{m-k}$ discrete Hilbert transform

© Springer Nature Switzerland AG 2019
E. Liflyand, *Functions of Bounded Variation and Their Fourier Transforms*,
Applied and Numerical Harmonic Analysis, https://doi.org/10.1007/978-3-030-04429-9

$\hbar_o a(m) = \sum_{\substack{k=1 \\ k \neq m}}^{\infty} \frac{2ka_k}{m^2-k^2} - \frac{a_m}{2m}$ \hfill odd discrete Hilbert transform

$\hbar_e a(m) = \sum_{\substack{k=1 \\ k \neq m}}^{\infty} \frac{2ma_k}{m^2-k^2} + \frac{a_m}{2m}$ \hfill even discrete Hilbert transform

$\int_0^{\frac{x}{2}} \frac{g(x-t)-g(x+t)}{t} dt = \int_{\frac{x}{2}}^{\frac{3x}{2}} \frac{g(t)}{x-t} dt$ \hfill T-transform of a function g

$L_0^1(\mathbb{R})$ \hfill space of wavelet functions

$L \log L = \{f : \int_{\mathbb{R}} |f(t)| \ln^+ |f(t)| \, dt < \infty\}$ \hfill Zygmund class

$\|g\|_{H^1(\mathbb{R})} = \|g\|_{L^1(\mathbb{R})} + \|\mathcal{H}g\|_{L^1(\mathbb{R})}$ \hfill norm of g in the real Hardy space

$H_o^1(\mathbb{R}_+)$ \hfill subspace of odd functions in the real Hardy space

$H_e^1(\mathbb{R}_+)$ \hfill subspace of even functions in the real Hardy space

$\|g\|_{O_q} = \int_0^\infty \left(\frac{1}{x} \int_x^{2x} |g(t)|^q dt \right)^{\frac{1}{q}} dx, \ 1 < q < \infty,$

$\|g\|_{O_\infty} = \int_0^\infty \operatorname*{ess\,sup}_{x \leq t \leq 2x} |g(t)| \, dx$ \hfill norms in the subspaces of $H_o^1(\mathbb{R}_+)$

$L^* := L^*(\mathbb{R}) = \{f : \|f\|_{L^*} = \int_0^\infty \operatorname*{ess\,sup}_{|t| \geq x} |f(t)| \, dx < \infty\}$

$E_q := E_q(\mathbb{R}_+) := \{g \in O_q : \int_0^\infty \frac{1}{x} \left| \int_0^x g(t) \, dt \right| dx < \infty\}$ \hfill subspaces of $H_e^1(\mathbb{R}_+)$

$\|g\|_{A_{1,2}} = \sum_{m=-\infty}^{\infty} \left\{ \sum_{j=1}^{\infty} \left[\int_{j2^m}^{(j+1)2^m} |g(t)| \, dt \right]^2 \right\}^{\frac{1}{2}} dx < \infty$ \hfill norm of g in $A_{1,2}$

$W(L^1, \ell^2)$ \hfill Wiener amalgam space

$B_\varphi g(x) = \frac{1}{x^2} \int_0^\infty g\left(\frac{t}{x}\right) \varphi(t) \, dt$ \hfill balance operator with kernel φ

$Q = \{g : g \in L^1(\mathbb{R}), \int_{\mathbb{R}} \frac{|\hat{g}(x)|}{|x|} dx < \infty\}$ \hfill widest space for the integrability of the
\hfill Fourier transform of a function of bounded variation

$Q_o = \{g : g \in L^1(\mathbb{R}), g(-t) = -g(t), \int_0^\infty \frac{|\hat{g}_s(x)|}{x} dx < \infty\}$ \hfill odd and

$Q_e = \{g : g \in L^1(\mathbb{R}), g(-t) = g(t), \int_0^\infty \frac{|\hat{g}_c(x)|}{x} dx < \infty\}$ \hfill even subspaces of Q

$H_Q^1(\mathbb{R}_+)$ \hfill consists of Q_o functions g with integrable $\mathcal{H}_o B_s g$

$\operatorname{Ci}(u) = -\int_u^\infty \frac{\cos t}{t} dt$ \hfill integral cosine

$\operatorname{Si}(u) = \int_0^u \frac{\sin t}{t} dt = \frac{\pi}{2} - \int_u^\infty \frac{\sin t}{t} dt$ \hfill integral sine

$\|g\|_{\Sigma_q} = \int_0^\infty \left(\frac{1}{x}\int_{x\le t\le 2x}|B_s g(t)|^q dt\right)^{\frac{1}{q}} dx, \ 1 < q < \infty$

$\|g\|_{\Sigma_\infty} = \int_0^\infty \operatorname*{ess\,sup}_{x\le t\le 2x} |B_s g(t)|\, dx$ $\qquad$ norms of g in intermediate spaces

$\widehat{f}_\gamma(x) = \int_0^\infty f(t)\cos(xt - \frac{\pi\gamma}{2})\, dt$ $\qquad$ union of two Fourier transforms: the cosine for $\gamma = 0$ and the sine for $\gamma = 1$

$V_0^*(\mathbb{R}) = \{f: f\in LAC(0,\infty),\ \lim_{|t|\to+\infty} f(t) = 0,$ $\qquad$ class of functions

$\|f\|_{V_0^*} = \|f'\|_{L^*} = \int_0^\infty \operatorname*{ess\,sup}_{|t|\ge x}|f'(t)|\, dx < \infty\}$ $\qquad$ with derivative in L^*

Lip 1, Lip α $\qquad$ Lipschitz classes

$\eta,\ \chi$ and ζ $\qquad$ indicator n-dimensional vectors with the entries either 0 or 1 only

x_η $\qquad$ the $|\eta|$-tuple consisting only of x_j such that $\eta_j = 1$

$dx_\eta := \prod_{j:\eta_j=1} dx_j$

$\Delta_{u_\eta} f(x) = \left(\prod_{j:\eta_j=1} \Delta_{u_j}\right) f(x),$ $\qquad$ partial difference

with $\Delta_{u_j} f(x) = f(x + u_j e_j) - f(x),$

$D^\eta f(x) = \left(\prod_{j:\eta_j=1} \frac{\partial}{\partial x_j}\right) f(x)$

$x^\alpha = x_1^{\alpha_1}\ldots x_n^{\alpha_n}$

$x_{-\eta}$ $\qquad$ the $|\eta|$-tuple consisting only of $\frac{1}{x_j}$ for j such that $\eta_j = 1$

$VV,\ VH,\ VT$ $\qquad$ Vitali, Hardy, and Tonelli variations

$\widehat{f}(x) = \int_{\mathbb{R}^n} f(u)e^{-i\langle x,u\rangle}\, du$ $\qquad$ multidimensional Fourier transform

$\langle x, u\rangle = x_1 u_1 + \cdots + x_n u_n$

J_ν $\qquad$ the Bessel function of first kind and order ν

$H^1(\mathbb{R}^n),\ H_m^1 = H^1(\mathbb{R}^{n_1}\times\ldots\times\mathbb{R}^{n_m}),$
$H^1(\mathbb{R}\times\ldots\times\mathbb{R}) := H_n^1(\mathbb{R}\times\ldots\times\mathbb{R})$ $\qquad$ real Hardy space and product Hardy spaces

$Q_H = \{g\in L^1(\mathbb{R}^n): \|g\|_{Q_H} = \|g\|_{L^1(\mathbb{R}^n)} + \int_{\mathbb{R}^n} \frac{|\widehat{g}(x)|}{|x_1\ldots x_n|}\, dx < \infty\}$ $\qquad$ widest space for the integrability of the Fourier transform of a VH function

$$\widehat{f}_\eta(x) = \int_{\mathbb{R}^n_+} f(u) \prod_{i:\eta_i=1} \cos x_i u_i \prod_{i:\eta_i=0} \sin x_i u_i \, du, \qquad \text{two forms of multidimensional}$$

$$\widehat{f}_\gamma(x) = \int_{\mathbb{R}^n_+} f(u) \left(\prod_{j=1}^n \cos\left(x_j u_j - \tfrac{\pi \gamma_j}{2}\right) \right) du \qquad \text{cosine-sine Fourier transform}$$

$MV^b_{\alpha+1}$ class of radial function, a substitute for functions of bounded variation

Index

Usually, page numbers show only the first appearance of the indicated term or name, except very few cases when the same term has different meaning in different appearances or is defined in a different setting.

Q space, 86
T-transform, 43

absolute continuity, 13, 128
Andersen, 51
Appel, xi
atomic characterization, 37

balance operator, 52
Banas, xi
Belinsky, 169
Bessel function, 123
Beurling, 43
Boas–Telyakovskii theorem, 3
Bochner, 2
Borwein, 43
Bownik, 37
bridge, 163
Butzer, 2

cancelation property, 28
Carroll, 8
Cauchy–Poisson formula, 123
characteristic (indicator) function, 17, 122
Coifman, 39
Cossar, 180
Cwikel, 3

Demeter, 8

even Hilbert transform, 30

F. and M. Riesz theorem, 78
Faddeev, 43
Flett, 43, 51
Fomin, 43
Fourier inversion, 124
Fourier multiplier, 19
Fourier transform, 16, 122
Fourier–Hardy inequality, 4, 11, 31
fractional Weyl derivative, 179
function of bounded variation, 12

Gaussian, 124
general monotone function, 61
Gokhberg, 36
Guberman, 3

Hankel transform, 122
Hardy, 51
Hardy variation, 120
Hausdorff–Young inequality, 18, 124
Helson, 79
Herz, 43
Hilbert transform, 19
Hilbert–Stieltjes transform, 23
Hille, 2

indicator vector, 117
inverse formula for the Hilbert transform, 100
Iosevich, 1
Ivashev-Musatov, 8

Johnson, 85

Khrushchev, 79
Kolmogorov example, 18
Kondurar, 15
Koosis, 79
Krause, 120
Krupnik, 36

Lebesgue decomposition, 14
Leray, 187
Littlewood, 51
Lowdenslager, 79

M. Riesz theorem, 36
Merentes Diáz, xi
modified Hilbert transform, 32
molecular characterization, 40
Mumford, 3
Muscalu, 8

Nessel, 2
Nevanlinna, 32

odd Hilbert transform, 30

Pólya, 1
Pólya theorem, 72
Paley–Wiener theorem, 50
Pichorides, 36
Plancherel, 18, 124
Poisson, 124
Poisson summation formula, 124
product Hardy space, 125

radial function, 122
Riemann–Lebesgue lemma, 17
Riemann–Liouville integral, 180
robust convexity, 146

Schlag, 8
Shevchuk, 47
Shilov, 2
singular function, 14
Stechkin, 70

Stieltjes integration, 15
Stirling numbers, 126
Sz.-Nagy, 2
Sz.-Nagy theorem, 111

Tamarkin, 2
Telyakovskii, 43
Tikhonov, 61
Titchmarsh, 21
Tonelli variation, 121
Trigub, xi, 3

Uchiyama, 127

Verkhovsky, xii
Vinogradov, 79
Vitali theorem, 27
Vitali variation, 119
Vysotsky, 16, 191

Warner, 85
weak estimate, 30
weak monotone function, 52
Weiss, 39
Whittaker–Kotel'nikov–Shannon sampling, 161
widest integrability space, 128
Wiener, 8, 49
Wiener algebra, 123
Wintner, 8

Zygmund–Stein condition, 33

Applied and Numerical Harmonic Analysis
(92 volumes)

1. A. I. Saichev and W.A. Woyczyński: *Distributions in the Physical and Engineering Sciences* (ISBN 978-0-8176-3924-2)

2. C.E. D'Attellis and E.M. Fernandez-Berdaguer: *Wavelet Theory and Harmonic Analysis in Applied Sciences* (ISBN 978-0-8176-3953-2)

3. H.G. Feichtinger and T. Strohmer: *Gabor Analysis and Algorithms* (ISBN 978-0-8176-3959-4)

4. R. Tolimieri and M. An: *Time-Frequency Representations* (ISBN 978-0-8176-3918-1)

5. T.M. Peters and J.C. Williams: *The Fourier Transform in Biomedical Engineering* (ISBN 978-0-8176-3941-9)

6. G.T. Herman: *Geometry of Digital Spaces* (ISBN 978-0-8176-3897-9)

7. Teolis: *Computational Signal Processing with Wavelets* (ISBN 978-0-8176-3909-9)

8. J. Ramanathan: *Methods of Applied Fourier Analysis* (ISBN 978-0-8176-3963-1)

9. J.M. Cooper: *Introduction to Partial Differential Equations with MATLAB* (ISBN 978-0-8176-3967-9)

10. Procházka, N.G. Kingsbury, P.J. Payner, and J. Uhlir: *Signal Analysis and Prediction* (ISBN 978-0-8176-4042-2)

11. W. Bray and C. Stanojevic: *Analysis of Divergence* (ISBN 978-1-4612-7467-4)

12. G.T. Herman and A. Kuba: *Discrete Tomography* (ISBN 978-0-8176-4101-6)

13. K. Gröchenig: *Foundations of Time-Frequency Analysis* (ISBN 978-0-8176-4022-4)

14. L. Debnath: *Wavelet Transforms and Time-Frequency Signal Analysis* (ISBN 978-0-8176-4104-7)

15. J.J. Benedetto and P.J.S.G. Ferreira: *Modern Sampling Theory* (ISBN 978-0-8176-4023-1)

16. D.F. Walnut: *An Introduction to Wavelet Analysis* (ISBN 978-0-8176-3962-4)

17. Abbate, C. DeCusatis, and P.K. Das: *Wavelets and Subbands* (ISBN 978-0-8176-4136-8)

18. O. Bratteli, P. Jorgensen, and B. Treadway: *Wavelets Through a Looking Glass* (ISBN 978-0-8176-4280-80

19. H.G. Feichtinger and T. Strohmer: *Advances in Gabor Analysis* (ISBN 978-0-8176-4239-6)

20. O. Christensen: *An Introduction to Frames and Riesz Bases* (ISBN 978-0-8176-4295-2)

21. L. Debnath: *Wavelets and Signal Processing* (ISBN 978-0-8176-4235-8)

22. G. Bi and Y. Zeng: *Transforms and Fast Algorithms for Signal Analysis and Representations* (ISBN 978-0-8176-4279-2)

23. J.H. Davis: *Methods of Applied Mathematics with a MATLAB Overview* (ISBN 978-0-8176-4331-7)

24. J.J. Benedetto and A.I. Zayed: *Sampling, Wavelets, and Tomography* (ISBN 978-0-8176-4304-1)

25. E. Prestini: *The Evolution of Applied Harmonic Analysis* (ISBN 978-0-8176-4125-2)

26. L. Brandolini, L. Colzani, A. Iosevich, and G. Travaglini: *Fourier Analysis and Convexity* (ISBN 978-0-8176-3263-2)

27. W. Freeden and V. Michel: *Multiscale Potential Theory* (ISBN 978-0-8176-4105-4)

28. O. Christensen and K.L. Christensen: *Approximation Theory* (ISBN 978-0-8176-3600-5)

29. O. Calin and D.-C. Chang: *Geometric Mechanics on Riemannian Manifolds* (ISBN 978-0-8176-4354-6)

30. J.A. Hogan: *Time–Frequency and Time–Scale Methods* (ISBN 978-0-8176-4276-1)

31. Heil: *Harmonic Analysis and Applications* (ISBN 978-0-8176-3778-1)

32. K. Borre, D.M. Akos, N. Bertelsen, P. Rinder, and S.H. Jensen: *A Software-Defined GPS and Galileo Receiver* (ISBN 978-0-8176-4390-4)

33. T. Qian, M.I. Vai, and Y. Xu: *Wavelet Analysis and Applications* (ISBN 978-3-7643-7777-9)

34. G.T. Herman and A. Kuba: *Advances in Discrete Tomography and Its Applications* (ISBN 978-0-8176-3614-2)

35. M.C. Fu, R.A. Jarrow, J.-Y. Yen, and R.J. Elliott: *Advances in Mathematical Finance* (ISBN 978-0-8176-4544-1)

36. O. Christensen: *Frames and Bases* (ISBN 978-0-8176-4677-6)

37. P.E.T. Jorgensen, J.D. Merrill, and J.A. Packer: *Representations, Wavelets, and Frames* (ISBN 978-0-8176-4682-0)

38. M. An, A.K. Brodzik, and R. Tolimieri: *Ideal Sequence Design in Time-Frequency Space* (ISBN 978-0-8176-4737-7)

39. S.G. Krantz: *Explorations in Harmonic Analysis* (ISBN 978-0-8176-4668-4)

40. Luong: *Fourier Analysis on Finite Abelian Groups* (ISBN 978-0-8176-4915-9)

41. G.S. Chirikjian: *Stochastic Models, Information Theory, and Lie Groups, Volume 1* (ISBN 978-0-8176-4802-2)

42. Cabrelli and J.L. Torrea: *Recent Developments in Real and Harmonic Analysis* (ISBN 978-0-8176-4531-1)

43. M.V. Wickerhauser: *Mathematics for Multimedia* (ISBN 978-0-8176-4879-4)

44. B. Forster, P. Massopust, O. Christensen, K. Gröchenig, D. Labate, P. Vandergheynst, G. Weiss, and Y. Wiaux: *Four Short Courses on Harmonic Analysis* (ISBN 978-0-8176-4890-9)

45. O. Christensen: *Functions, Spaces, and Expansions* (ISBN 978-0-8176-4979-1)

46. J. Barral and S. Seuret: *Recent Developments in Fractals and Related Fields* (ISBN 978-0-8176-4887-9)

47. O. Calin, D.-C. Chang, and K. Furutani, and C. Iwasaki: *Heat Kernels for Elliptic and Sub-elliptic Operators* (ISBN 978-0-8176-4994-4)

48. C. Heil: *A Basis Theory Primer* (ISBN 978-0-8176-4686-8)

49. J.R. Klauder: *A Modern Approach to Functional Integration* (ISBN 978-0-8176-4790-2)

50. J. Cohen and A.I. Zayed: *Wavelets and Multiscale Analysis* (ISBN 978-0-8176-8094-7)

51. Joyner and J.-L. Kim: *Selected Unsolved Problems in Coding Theory* (ISBN 978-0-8176-8255-2)

52. G.S. Chirikjian: *Stochastic Models, Information Theory, and Lie Groups, Volume 2* (ISBN 978-0-8176-4943-2)

53. J.A. Hogan and J.D. Lakey: *Duration and Bandwidth Limiting* (ISBN 978-0-8176-8306-1)

54. Kutyniok and D. Labate: *Shearlets* (ISBN 978-0-8176-8315-3)

55. P.G. Casazza and P. Kutyniok: *Finite Frames* (ISBN 978-0-8176-8372-6)

56. V. Michel: *Lectures on Constructive Approximation* (ISBN 978-0-8176-8402-0)

57. D. Mitrea, I. Mitrea, M. Mitrea, and S. Monniaux: *Groupoid Metrization Theory* (ISBN 978-0-8176-8396-2)

58. T.D. Andrews, R. Balan, J.J. Benedetto, W. Czaja, and K.A. Okoudjou: *Excursions in Harmonic Analysis, Volume 1* (ISBN 978-0-8176-8375-7)

59. T.D. Andrews, R. Balan, J.J. Benedetto, W. Czaja, and K.A. Okoudjou: *Excursions in Harmonic Analysis, Volume 2* (ISBN 978-0-8176-8378-8)

60. D.V. Cruz-Uribe and A. Fiorenza: *Variable Lebesgue Spaces* (ISBN 978-3-0348-0547-6)

61. W. Freeden and M. Gutting: *Special Functions of Mathematical (Geo-)Physics* (ISBN 978-3-0348-0562-9)

62. A. I. Saichev and W.A. Woyczyński: *Distributions in the Physical and Engineering Sciences, Volume 2: Linear and Nonlinear Dynamics of Continuous Media* (ISBN 978-0-8176-3942-6)

63. S. Foucart and H. Rauhut: *A Mathematical Introduction to Compressive Sensing* (ISBN 978-0-8176-4947-0)

64. Herman and J. Frank: *Computational Methods for Three-Dimensional Microscopy Reconstruction* (ISBN 978-1-4614-9520-8)

65. Paprotny and M. Thess: *Realtime Data Mining: Self-Learning Techniques for Recommendation Engines* (ISBN 978-3-319-01320-6)

66. Zayed and G. Schmeisser: *New Perspectives on Approximation and Sampling Theory: Festschrift in Honor of Paul Butzer's 85th Birthday* (ISBN 978-3-319-08800-6)

67. R. Balan, M. Begue, J. Benedetto, W. Czaja, and K.A Okoudjou: *Excursions in Harmonic Analysis, Volume 3* (ISBN 978-3-319-13229-7)

68. Boche, R. Calderbank, G. Kutyniok, J. Vybiral: *Compressed Sensing and its Applications* (ISBN 978-3-319-16041-2)

69. S. Dahlke, F. De Mari, P. Grohs, and D. Labate: *Harmonic and Applied Analysis: From Groups to Signals* (ISBN 978-3-319-18862-1)

70. Aldroubi, *New Trends in Applied Harmonic Analysis* (ISBN 978-3-319-27871-1)

71. M. Ruzhansky: *Methods of Fourier Analysis and Approximation Theory* (ISBN 978-3-319-27465-2)

72. G. Pfander: *Sampling Theory, a Renaissance* (ISBN 978-3-319-19748-7)

73. R. Balan, M. Begue, J. Benedetto, W. Czaja, and K.A Okoudjou: *Excursions in Harmonic Analysis, Volume 4* (ISBN 978-3-319-20187-0)

74. O. Christensen: *An Introduction to Frames and Riesz Bases, Second Edition* (ISBN 978-3-319-25611-5)

75. E. Prestini: *The Evolution of Applied Harmonic Analysis: Models of the Real World, Second Edition* (ISBN 978-1-4899-7987-2)

76. J.H. Davis: *Methods of Applied Mathematics with a Software Overview, Second Edition* (ISBN 978-3-319-43369-1)

77. M. Gilman, E. M. Smith, S. M. Tsynkov: *Transionospheric Synthetic Aperture Imaging* (ISBN 978-3-319-52125-1)

78. S. Chanillo, B. Franchi, G. Lu, C. Perez, E.T. Sawyer: *Harmonic Analysis, Partial Differential Equations and Applications* (ISBN 978-3-319-52741-3)

79. R. Balan, J. Benedetto, W. Czaja, M. Dellatorre, and K.A Okoudjou: *Excursions in Harmonic Analysis, Volume 5* (ISBN 978-3-319-54710-7)

80. Pesenson, Q.T. Le Gia, A. Mayeli, H. Mhaskar, D.X. Zhou: *Frames and Other Bases in Abstract and Function Spaces: Novel Methods in Harmonic Analysis, Volume 1* (ISBN 978-3-319-55549-2)

81. Pesenson, Q.T. Le Gia, A. Mayeli, H. Mhaskar, D.X. Zhou: *Recent Applications of Harmonic Analysis to Function Spaces, Differential Equations, and Data Science: Novel Methods in Harmonic Analysis, Volume 2* (ISBN 978-3-319-55555-3)

82. F. Weisz: *Convergence and Summability of Fourier Transforms and Hardy Spaces* (ISBN 978-3-319-56813-3)

83. Heil: *Metrics, Norms, Inner Products, and Operator Theory* (ISBN 978-3-319-65321-1)

84. S. Waldron: *An Introduction to Finite Tight Frames: Theory and Applications.* (ISBN: 978-0-8176-4814-5)

85. Joyner and C.G. Melles: *Adventures in Graph Theory: A Bridge to Advanced Mathematics.* (ISBN: 978-3-319-68381-2)

86. B. Han: *Framelets and Wavelets: Algorithms, Analysis, and Applications* (ISBN: 978-3-319-68529-8)

87. H. Boche, G. Caire, R. Calderbank, M. März, G. Kutyniok, R. Mathar: *Compressed Sensing and Its Applications* (ISBN: 978-3-319-69801-4)

88. N. Minh Chong: *Pseudodifferential Operators and Wavelets over Real and p-adic Fields* (ISBN: 978-3-319-77472-5)

89. A. I. Saichev and W.A. Woyczyński: *Distributions in the Physical and Engineering Sciences, Volume 3: Random and Fractal Signals and Fields* (ISBN: 978-3-319-92584-4)

90. K. Bredies and D. Lorenz: *Mathematical Image Processing* (ISBN: 978-3-030-01457-5)

91. Boggiatto, P., Cordero, E., de Gosson, M., Feichtinger, H.G., Nicola, F., Oliaro, A., Tabacco, A.: Landscapes of Time-Frequency Analysis (ISBN: 978-3-030-05209-6)
92. Liflyand, E: Functions of Bounded Variation and Their Fourier Transforms (ISBN: 978-3-030-04428-2)

For an up-to-date list of ANHA titles, please visit http://www.springer.com/series/4968

GPSR Compliance
The European Union's (EU) General Product Safety Regulation (GPSR) is a set of rules that requires consumer products to be safe and our obligations to ensure this.

If you have any concerns about our products, you can contact us on

ProductSafety@springernature.com

In case Publisher is established outside the EU, the EU authorized representative is:

Springer Nature Customer Service Center GmbH
Europaplatz 3
69115 Heidelberg, Germany

www.ingramcontent.com/pod-product-compliance
Ingram Content Group UK Ltd.
Pitfield, Milton Keynes, MK11 3LW, UK
UKHW021252180426
11946UKWH00004B/97